Kinetic Systems

KINETIC SYSTEMS

Mathematical Description
of Chemical Kinetics in Solution

CHRISTOS CAPELLOS
Feltman Research Laboratory
Picatinny Arsenal, Dover New Jersey

BENON H. J. BIELSKI
Brookhaven National Laboratory
Upton, New York

WILEY-INTERSCIENCE,
a Division of John Wiley & Sons, Inc.
New York . London . Sydney . Toronto

Copyright © 1972, by John Wiley & Sons, Inc.

All rights reserved. Published simultaneously in Canada.

No part of this book may be reproduced by any means, nor transmitted, nor translated into a machine language without the written permission of the publisher.

Library of Congress Cataloging in Publication Data

Capellos, Christos.
  Kinetic systems.

    Includes bibliographies.
    1. Chemical reaction, Rate of—Mathematical models. I.   Bielski, Benon H. J., joint author. II. Title.

QD501.C373   541′.39′0184   75-39723
ISBN 0-471-13450-3

Printed in the United States of America.

10 9 8 7 6 5 4 3 2 1

Preface

This book has been written for the chemistry student who wishes to become fluent in the mathematical operations encountered in chemical kinetics and for the researcher who needs a practical guidebook. Being primarily directed toward the student, the systematic step-by-step development of the mathematical equations is presented intentionally in an elementary fashion to keep the text at a self-explanatory level—the main purpose of this book. We hope that such presentation will enable the reader to follow the mathematical operations at a rapid pace by minimizing his consultation of other texts.

The material in this volume is restricted to chemical kinetics of reactions in solutions. It does not deal with the theory of chemical kinetics, since this subject has been very well treated in other books; instead it is devoted to a detailed and rigorous mathematical treatment of chemical kinetic systems. The chemical systems, some of which are discussed in detail in existing texts and some of which are of little practical use, serve here as a vehicle for a systematic development of the mathematical operations only.

Warm thanks are due to Dr. A. O. Allen of Brookhaven National Laboratory and Dr. R. F. Walker of F. R. L., Picatinny Arsenal who were most helpful in clarification and simplification of the text. Particular thanks go to Dr. J. M. Gebicki of Macquarie University, Australia and the publishers and their staff for their excellent help in the preparation of this volume.

CHRISTOS CAPELLOS
BENON H. J. BIELSKI

Contents

1. **Introduction** — 1
2. **Determination of Order of Reaction** — 4
3. **Zero-Order Reactions**
   System 1. $A \xrightarrow{k}$ product(s) — 7
4. **First-Order Reactions**
   System 2. $A \xrightarrow{k}$ product(s) — 9
5. **Second-Order Reactions**
   System 3. $A + B \xrightarrow{k}$ product(s)
   $[A]_0 \neq [B]_0$ — 14
   System 4. $A + B \xrightarrow{k}$ product(s)
   $[A]_0 = [B]_0$ — 16
   System 5. $A + A \xrightarrow{k}$ product(s) — 17
6. **Third-Order Reactions** $A + B + C \xrightarrow{k}$ **product(s)**
   System 6. $[A]_0 = [B]_0 = [C]_0$ — 19
   System 7. $[A]_0 = [B]_0 \neq [C]_0$ — 20
   System 8. $[A]_0 \neq [B]_0 \neq [C]_0$ — 24
7. **Reaction of nth Order**
   System 9. $nA \xrightarrow{k}$ product(s) — 28

viii Contents

8. Reversible Reactions — 30

 System 10. First-Order $A \underset{k_2}{\overset{k_1}{\rightleftharpoons}} B$

  Case I.   At $t = 0$, $[A] = [A]_0$, $[B] = 0$

   Classical Solution — 31

   Operator Method — 32

  Case II.   At $t = 0$, $[A] = [A]_0$, $[B] = [B]_0$ — 33

  Case III. System at Equilibrium — 34

 System 11. First-Order two-stage

  $A \underset{k_2}{\overset{k_1}{\rightleftharpoons}} B \underset{k_4}{\overset{k_3}{\rightleftharpoons}} C$ — 35

 System 12. Second-Order

  $A + B \underset{k_2}{\overset{k_1}{\rightleftharpoons}} C + D$ — 38

 System 13. First- and Second-Order

  $A \underset{k_2}{\overset{k_1}{\rightleftharpoons}} B + C$ — 41

 System 14. Second- and First-Order

  $A + B \underset{k_2}{\overset{k_1}{\rightleftharpoons}} C$ — 43

9. Consecutive Irreversible Reactions

 System 15. First-Order three-stage

  $A \xrightarrow{k_1} B \xrightarrow{k_2} C \xrightarrow{k_3} D$

  Classical Method — 46

  Operator Method — 49

 System 16. First-Order, $(n - 1)$-Stages

  $A_1 \xrightarrow{k_1} A_2 \xrightarrow{k_2} A_3 \xrightarrow{} \cdots A_{n-1} \xrightarrow{k_{n-1}} A_n$

  Classical Solution — 53

  Operator Method — 57

Contents ix

10. Catalytic Reactions

System 17. Second-Order Autocatalytic Reaction in Which One of the Products Acts as the Catalyst   59

System 18. Second-Order Autocatalytic Reaction in Which the Initially Added Catalyst Differs from the Reaction Product Catalyst   62

System 19. An Acid-Base Catalyzed Decay Reaction of First-Order   64

System 20. Second-Order Acid-Catalyzed Decay of a Dissociable Free Radical   65

11. Parallel Reactions

System 21. First-Order with Common Product

$$A \xrightarrow{k_1} C$$
$$B \xrightarrow{k_1} C$$
    68

System 22. First-Order with Different Products

$$A \xrightarrow{k_1} B$$
$$A \xrightarrow{k_2} C$$
$$A \xrightarrow{k_3} D$$
    69

System 23. Pseudo First-Order

$$A + C \xrightarrow{k_1} \text{product(s)}$$
$$B + C \xrightarrow{k_2} \text{product(s)}$$
$$[A], [B] \ll [C]$$
    72

System 24. First-Order Reversible and Irreversible

$$A \underset{k_2}{\overset{k_1}{\rightleftharpoons}} B$$
$$A \xrightarrow{k_3} C$$
    73

Contents

System 25. Second-Order
$$A + B \xrightarrow{k_1} \text{product(s)}$$
$$A + B \xrightarrow{k_2} \text{product(s)}$$
$$A + B \xrightarrow{k_3} \text{product(s)} \quad \quad 75$$
$$2k_1 = 2k_3 = k_2$$

System 26. Second-Order
$$A + B \xrightarrow{k_1} C$$
$$A + B \xrightarrow{k_2} D$$
$$A + B \xrightarrow{k_3} E \quad \quad 78$$

System 27. Second-Order
$$A \xrightarrow{k_1} B^- + C^+$$
$$C^+ + B^- \xrightarrow{k_2} \text{product(s)}$$
$$C^+ + D \xrightarrow{k_3} D^+ + \text{product(s)}$$
$$B^- + D^+ \xrightarrow{k_4} \text{product(s)} \quad \quad 80$$

System 28. First- and Second-Order
$$A \xrightarrow{k_1} \text{product(s)}$$
$$A + A \xrightarrow{k_2} \text{product(s)} \quad \quad 85$$

System 29. First- and Second-Order
$$A \xrightarrow{k_1} D + E$$
$$E + B \xrightarrow{k_2} C$$
$$A + B \xrightarrow{k_3} C + D$$
$$k_2 \gg k_1, k_3 \quad \quad 87$$

System 30. Pseudo First- and Second-Order
$$A + B \xrightarrow{k_1} C$$
$$A + A \xrightarrow{k_2} \text{product(s)} \quad \quad 89$$

Contents　xi

12. Consecutive Reactions of Higher Order　93

System 31. First- Followed by Second-Order

$$A \xrightarrow{k_1} B$$
$$A + B \xrightarrow{k_2} C \qquad 93$$

System 32. Second- Followed by First-Order

$$A + A \xrightarrow{k_1} B$$
$$B \xrightarrow{k_2} C \qquad 95$$

System 33. Second- Followed by Second-Order

$$A + B \xrightarrow{k_1} C$$
$$B + C \xrightarrow{k_2} D \qquad 100$$

13. Miscellaneous

System 34. Electron Transfer Reactions in Polar Solvents　104

System 35. Diffusion Controlled Reactions in Solutions　106

System 36. Effect of Coulombic Interactions on Reaction Velocities　107

System 37. Computation of the Extent of Decay of Transient during the Period (Electron Pulse) of its Formation　110

System 38. Effect of Temperature on Reaction Velocity　113

System 39. Effect of Pressure on Reaction Velocity　116

System 40. Effect of Ionic Strength on Velocity of Ionic Reactions　119

Problems　125

Appendix I	Differentials	**131**
	Indefinite Integrals	**132**
	Definite Integrals	**133**
	Definite Integrals of Some Special Functions	**133**
Appendix II	Guide Line to the Operator Method	**134**
	Table of Transforms and Originals	**137**

Kinetic Systems

1

Introduction

Chemical reactions are processes in which a substance or substances (reactants) are transformed into other substances (products). In some processes the change occurs directly and the complete description of the mechanism of the reaction presents few difficulties. However complex processes in which the substances undergo a series of stepwise changes, each constituting a reaction in its own right, are much more common. The overall mechanism is then made up of contributions from all such reactions, and it is usually far too complex to determine from the knowledge of the reactants and products alone. In these complex cases chemical kinetics can often provide the only feasible approach toward the unraveling of the reaction mechanism.

Chemical kinetics is concerned with the analysis of the dynamics of chemical reactions. The raw data of chemical kinetics are the measurements of rates of reactions; the desired end product is the explanation of these rates in terms of a complete reaction mechanism. It is obvious that since a measured rate reflects a statistical average state of the molecules taking part in the reaction, chemical kinetics provide no information on the energetic or stereochemical state of the individual molecules. However, it does have the valuable potential to break down complex mechanisms into sequences of simple reactions.

2 Introduction

This monograph examines a series of processes (systems) of increasing complexity, from the point of view of their kinetics. Model reactions are set up and the appropriate rate equations written. The equations are then used to arrive at an expression relating measurable parameters of the reactions to constants and to concentration terms that can be evaluated by graphical or numerical solutions from this relationship. When an exact expression cannot be derived, approximate solutions can often be found by using some simplifying assumptions. If the kinetics of a complex process are then found to fit closely with the model equation thus derived, the model can be used as a basis for description of the process. Further refinement and elimination of approximations may lead to considerable modifications of the description but a starting point for the determination of a mechanism can be provided by the model.

The parameters of interest in kinetics are the quantities of reactants and products and their rates of change. Since reactants disappear in reactions, their rate expressions are given a negative sign. The amounts of products increase and their rates of change are therefore positive. As they are seldom constant, the rates are written as differentials. Thus in a general reaction

$$aA + bB + \cdots \xrightarrow{k} cC + dD + \cdots$$

the reaction rates for the individual components are

$$-\frac{1}{a}\frac{d[A]}{dt} \; ; \quad -\frac{1}{b}\frac{d[B]}{dt} \; ; \quad +\frac{1}{c}\frac{d[C]}{dt} \; ; \quad +\frac{1}{d}\frac{d[D]}{dt}$$

The square brackets denote concentrations in moles per liter, denoted by M. Concentrations rather than amounts are used to make the rates independent of the volume of system. In addition, unless otherwise indicated, only closed, homogeneous systems are considered in which there is no gain or loss of material during the reaction. Reactions are assumed to proceed isothermally so that temperature can be treated as an independent variable.

Unit of time used is the second. The instant of time is tied to the concentration of any particular substance by the use of subscripts, for example, at time t_1 the concentration of A is $[A]_1$, at t_2 it is $[A]_2$, and so on. Terms such as $[A]_0$ and $t = 0$ refer to concentration of substance A at the instant before any A takes part in the reaction.

The rate of reaction at a fixed temperature is proportional to concentration terms,

$$-\frac{1}{a}\frac{d[A]}{dt} = k([A]^\alpha + [B]^\beta + \cdots)$$

The proportionality constant k is called the rate constant. Unit of k can be deduced from examination of the rate expression; it has dimensions of (concentration)$^{1-n}$(time)$^{-1}$.

The sum of the exponents of the concentrations $n = \alpha + \beta + \cdots$ is the overall order of the reaction, while α is the order of the reaction with respect to A, and β is the order of the reaction with respect to B. The order of reaction for each reacting compound must be determined experimentally since it cannot be deduced or predicted from the equation describing the reaction. The exponents may be positive integers or fractions (negative exponents have been also reported) and do not have to be equal to the stoichiometric coefficients of the reactants in the net reaction.

REFERENCES

1. E. A. Moelwyn-Hughes, *The Kinetics of Reactions in Solutions*, Oxford University Press, Oxford, 1933.
2. S. W. Benson, *The Foundations of Chemical Kinetics*, McGraw-Hill Book Company, Inc., New York, 1960.
3. A. A. Frost and R. G. Pearson, *Kinetics and Mechanism*, John Wiley and Sons, Inc., New York, 1961.
4. H. Eyring and E. M. Eyring, *Modern Chemical Kinetics*, Reinhold Publishing Corporation, New York, 1963.
5. M. Ritchie, *Chemical Kinetics in Homogeneous Systems*, John Wiley and Sons, Inc., New York, 1966.

2

Determination of Order of Reaction

THE HALF-LIFE METHOD

The time required for one-half of the initial concentration of a given reactant to be consumed is called the half-life, $t_{1/2}$. Since $t_{1/2}$ depends on the order of reaction, its determination at various initial concentrations can be used to evaluate the order. Application of this method is simple for reactions of only one reactant but becomes quite tedious for multiple component systems.

For reactions of a single species the relationship between the half-life and the respective reaction orders are as follows.

Order of Reaction	Experimental Straight Line Plots	Half-Life, $t_{1/2}$
Zero	$[A]$ vs. time	$[A]_0/2k$
First	$\ln [A]$ vs. time	$(1/k) \ln 2 = 0.693/k$
Second	$1/[A]$ vs. time	$1/(k[A]_0)$
Third	$1/[A]^2$ vs. time	$3/2k[A]_0^2$

SUBSTITUTION METHOD

This method is applicable to reactions that are not complex and fit either zero-, first-, or second-order kinetics. By substituting data into the proper equation the computed value for k should stay constant throughout the course of the reaction.

THE INITIAL RATE METHOD

For a reaction of the nth order the rate of reaction is given by

$$R_1 = \frac{d[A]_1}{dt} = k_n[A]_1^n \qquad (1)$$

where the subscript 1 refers to a particular initial rate. The order of reaction can be determined from measurements of initial rates R_1 and R_2 for different initial concentrations of A, since

$$R_1 = k_n[A]_1^n \quad \text{and} \quad R_2 = k_n[A]_2^n$$

and

$$n = \left[\frac{\ln(R_1/R_2)}{\ln[A]_1/[A]_2}\right] \qquad (2)$$

THE ISOLATION METHOD

In a second-order reaction when one of the reactants is present in large excess, the reaction will behave experimentally as if it were of first-order. Similarly a third-order reaction behaves as if it were second-order if one reactant is in large excess, or it behaves as a first-order reaction if two reactants are in excess. The reason for such behavior (pseudoorder reactions) is that the concentration of the species present in large excess will remain virtually constant during the course of the chemical change, and the overall order of the reaction will be apparently reduced.

6 Determination of Order of Reaction

In principle it is possible to isolate each of the reacting species by adjustment of the concentrations of the other participating compounds. Ostwald[1] utilized this technique for the determination of the order of reactions, as illustrated with the following general example:

$$\frac{dx}{dt} = k([A]_0 - x)^a([B]_0 - x)^b([C]_0 - x)^c \tag{3}$$

Here a, b, and c are the unknown exponents. The individual exponents are determined by choice of the proper initial concentrations, for example, if

$$[A]_0 \quad \text{and} \quad [B]_0 \gg [C]_0$$

equation 3 can be written

$$\frac{dx}{dt} = k[A]_0{}^a[B]_0{}^b([C]_0 - x)^c \tag{4}$$

and the reaction is of order c, since the product $k[A]_0{}^a[B]_0{}^b$ is constant.

If $[A]_0$ and $[C]_0 \gg [B]_0$, equation 3 becomes

$$\frac{dx}{dt} = k[A]_0{}^a[C]_0{}^c([B]_0 - x)^b \tag{5}$$

and reaction is of order b; similarly if $[B]_0$ and $[C]_0 \gg [A]_0$ the observed order would be a. For experimental conditions where $[A]_0 = [B]_0 = [C]_0$ the order of the observed reaction is $n = a + b + c$.

REFERENCE

1. W. Ostwald, *Lehrbuch der Allgemeinen Chemie*, Vol. 2, Leipzig, 1887.

3

Zero-order Reactions

SYSTEM 1

The rate of a chemical reaction is of zero-order if it is independent of the concentrations of the participating substances. In reactions of this type the rate of reaction is determined by such limiting factors as follows.

1. In catalysed processes the rates of diffusion of reactants, availability of surface sites, and such.
2. In photochemistry, the intensity and wave length of light.
3. In radiation chemistry the energy, intensity and nature of radiation.

If the rate of the reaction is independent of the concentration of the reacting substance A, then the amount $d[A]$ by which the concentration of A decreases in any given unit of time dt is constant throughout the course of the reaction:

$$-\frac{d[A]}{dt} = k_1 \tag{1-1}$$

By convention, the negative sign shows that in the particular reaction described, A is removed from the system and k_1 is the velocity constant whose units are conventionally moles per liter per second ($M\ \text{sec}^{-1}$). Assuming that at time t_1, the concentration of A (denoted by square brackets and normally

8 Zero-order Reactions

expressed in moles per liter) is $[A]_0$ and at time t_2 the concentration is $[A]$, equation 1-1 can be integrated between these limits to give

$$-\int_{[A]_0}^{[A]} d[A] = k_1 \int_{t_1}^{t_2} dt \qquad (1\text{-}2)$$

$$-[A]\Big|_{[A]_0}^{[A]} = k_1 t \Big|_{t_1}^{t_2} \qquad (1\text{-}3)$$

$$\begin{array}{c} |-[A] = k_1 t_2 + I| \\ -|-[A]_0 = k_1 t_1 + I| \\ \hline [A] - [A]_0 = -k_1(t_2 - t_1) \end{array} \quad (I = \text{integration constant}) \qquad (1\text{-}4)$$

hence

$$k_1 = \frac{[A]_0 - [A]}{(t_2 - t_1)} \qquad (1\text{-}5)$$

If $t_1 = 0$, equation 1-5 reduces to

$$[A] = [A]_0 - k_1 t_2 \qquad (1\text{-}6)$$

and a plot of $[A]$ versus time yields a straight line where $[A]_0$ is the intercept and k_1 the slope.

The velocity constant k_1 may include arbitrary constants arising from various limiting factors such as diffusion constants and a fixed intensity of absorbed light.

A system in which several reactants undergo simultaneously zero-order reactions can be described by the following general equation:

$$aA + bB + cC \rightarrow \text{products}$$

It is possible to write a rate expression similar to equation 1-1 for each reactant and to show that the ratios of the specific rate constants to the integral coefficients (a, b, c) are equal:

$$\text{Rate} = -\frac{1}{a}\frac{d[A]}{dt} = -\frac{1}{b}\frac{d[B]}{dt} = -\frac{1}{c}\frac{d[C]}{dt} \qquad (1\text{-}7)$$

so that

$$\frac{k_A}{a} = \frac{k_B}{b} = \frac{k_C}{c} = k \qquad (1\text{-}8)$$

4

First-order Reactions

SYSTEM 2

$$A \xrightarrow{k_1} B \qquad \text{(I)}$$

The rate of a first-order reaction is proportional to the first power of the concentration of only one reactant. This means that the amount $d[A]$, which undergoes chemical change in the short time interval dt, depends only on the amount of A present at that instant, assuming that there is no change in volume, temperature, or any other factor that may affect the reaction.

The rate expression which describes a first-order reaction is

$$-\frac{d[A]}{dt} = k_1[A] \qquad (2\text{-}1)$$

Equation 2-1 can be rearranged and integrated as an indefinite integral to give:

$$-\int \frac{d[A]}{[A]} = k_1 \int dt \qquad (2\text{-}2)$$

$$-\ln[A] = k_1 t + I \qquad (2\text{-}3)$$

where I is the integration constant determined by the experimental conditions. The evaluation of I is discussed later in this section.

10 First-order Reactions

Upon integrating (2-2) between limits one obtains for each limiting condition an equation equivalent to (2-3). These equations can be subtracted from each other with the resultant elimination of the integration constant:

$$-\int_{[A]_0}^{[A]} \frac{d[A]}{[A]} = k_1 \int_{t_1}^{t_2} dt \qquad (2\text{-}4)$$

$$\frac{\begin{array}{c}|-\ln [A] = k_1 t_2 + I|\\ -|-\ln [A]_0 = k_1 t_1 + I|\end{array}}{-\ln [A] + \ln [A]_0 = k_1(t_2 - t_1)} \qquad (2\text{-}5)$$

$$\ln \frac{[A]_0}{[A]} = k_1(t_2 - t_1) \qquad (2\text{-}6)$$

The left-hand side of equation 2-6 can be converted to the decadic logarithm by multiplying it by 2.303. The exponential form of this equation is

$$[A] = [A]_0 e^{-k_1(t_2 - t_1)} \qquad (2\text{-}7)$$

A modified version of the above derivation defines $[A]_0$ as the initial quantity of the reacting material A in a given volume, and x as the amount which reacts in time t, hence $([A]_0 - x)$ is the amount of A remaining after time t.

$$A \xrightarrow{k_1} B \qquad (\text{II})$$

At time $t = 0$ $\qquad [A]_0 \qquad\qquad 0$
At time $t = t$ $\qquad ([A]_0 - x) \qquad x$

For this formulation the rate of formation of product is given by the following linear differential equation:

$$\frac{dx}{dt} = k_1([A]_0 - x) \qquad (2\text{-}8)$$

which has the general form

$$\frac{dx}{dt} + k_1 x = k_1[A]_0 \qquad (2\text{-}9)$$

and can be integrated directly:

$$\int \frac{dx}{([A]_0 - x)} = k_1 \int dt \tag{2-10}$$

$$-\int \frac{d([A]_0 - x)}{([A]_0 - x)} = k_1 \int dt \tag{2-11}$$

$$-\ln([A]_0 - x) + I = k_1 t \tag{2-12}$$

The integration constant I is evaluated from the boundary conditions. At time $t = 0$, $x = 0$; hence equation 2-12 reduces to

$$-\ln[A]_0 + I = 0 \tag{2-13}$$

or

$$\ln[A]_0 = I \tag{2-14}$$

Substitution of (2-14) into (2-12) yields

$$-\ln([A]_0 - x) + \ln[A]_0 = k_1 t \tag{2-15}$$

or

$$k_1 = \frac{1}{t} \ln \frac{[A]_0}{([A]_0 - x)} \tag{2-16}$$

The exponential form of (2-16) is

$$x = [A]_0 (1 - e^{-k_1 t}) \tag{2-17}$$

This equation expresses the concentration of the product B at any time t in terms of the initial concentration of A and the rate constant k. It is apparent that when $t = 0$, $e^{-k_1 t} = 1$, hence $x = 0$. With increasing t, $e^{-k_1 t}$ at first decreases rapidly, then slowly approaches zero. Hence x approaches asymptotically the value of $[A]_0$, which corresponds to complete conversion of A into B.

The unit of the rate constant for a first-order reaction is a number per unit of time expressed in reciprocal seconds, for example, $k = 0.123$ sec^{-1}. Instead of the rate constant a quantity referred to as the half-life is often used, which is designated by the symbol $t_{1/2}$. This quantity is more tangible since it tells how much time is required for the concentration of

First-order Reactions

A to fall to one-half of its value from some initial measurement. Substitution of the appropriate numerical values into equation 2-6 yields

$$k = \frac{2.303}{t_{1/2}} \log \frac{(1)}{(\frac{1}{2})} = \frac{0.693}{t_{1/2}} \qquad (2\text{-}18)$$

It is important to note that this equation shows that in first order reactions the half-life is independent of the concentration of A. Often this can be used to test whether a reaction obeys first-order kinetics by measuring half-lives of the reaction at various initial concentrations of A.

Another quantity often encountered is the *mean life* of this reaction, designated by the symbol τ. It is by definition the time required for the concentration of $[A]_0$ to fall to $[A]_0/e$, where $e = 2.718$, or

$$\tau = \frac{1}{k} \qquad (2\text{-}19)$$

First-order reactions often occur simultaneously with reactions of other orders as part of an overall mechanism. As the mathematical manipulations of kinetic equations are simplest for first-order reactions, the conditions of more complex chemical processes are sometimes adjusted so that the measured overall rate becomes dependent on a single concentration term. The process is then said to be of pseudo first-order.

REFERENCES

1. L. Whilhelmy, *Pogg. Annalen*, **81**, 413 (1850).
2. T. M. Lowry, *Trans. Chem. Soc. (London)*, **75**, 212 (1899).
3. C. S. Hudson and J. K. Dale, *J. Am. Chem. Soc.*, **39**, 320 (1917).
4. F. Daniels and E. H. Johnston, *J. Am. Chem. Soc.*, **43**, 53 (1921).
5. H. Eyring and F. Daniels, *J. Am. Chem. Soc.*, **52**, 1472 (1930).
6. E. A. Moelwyn-Hughes, *Z. Phys. Chem.*, **B26**, 281 (1934).
7. R. A. Ogg, *J. Chem. Phys.*, **15**, 337 (1947).

8. R. A. Ogg, *J. Chem. Phys.*, **15,** 613 (1947).
9. R. A. Ogg, *J. Chem. Phys.*, **18,** 572 (1950).
10. G. Czapski and B. H. J. Bielski, *J. Phys. Chem.*, **67,** 2180 (1963).
11. C. Capellos and A. O. Allen, *J. Phys. Chem.*, **72,** 4265 (1968).
12. B. H. J. Bielski, *J. Phys. Chem.*, **74,** 3213 (1970).
13. C. Capellos and A. O. Allen, *J. Phys. Chem.*, **74,** 840 (1970).

5

Second-order Reactions

SYSTEM 3. TYPE I

When two reactants A and B interact in such a way that the rate of reaction is proportional to the first power of the product of their respective concentrations, the compounds are said to undergo a second-order reaction.

If $[A]_0$ and $[B]_0$ designate the initial quantities of the two reacting chemicals A and B, and x is the number of moles of A or B which react in a given time interval t, then the rate of formation of product C can be described by the following mechanism and set of equations:

$$A \quad + \quad B \quad \xrightarrow{k_1} C \qquad \text{(I)}$$

Amount at time $t = 0$ $\quad [A]_0 \quad\quad [B]_0 \quad\quad 0$
Amount at time $t = t$ $\quad ([A]_0 - x) \quad ([B]_0 - x) \quad x$

Rate of reaction is then,

$$\frac{dx}{dt} = k_1([A]_0 - x)([B]_0 - x) \qquad (3\text{-}1)$$

or

$$\frac{dx}{([A]_0 - x)([B]_0 - x)} = k_1\, dt \qquad (3\text{-}2)$$

Equation 3-2 can be integrated after resolution into partial fractions:

$$\frac{1}{([A]_0 - x)([B]_0 - x)} = \frac{p}{([A]_0 - x)} + \frac{q}{([B]_0 - x)} \quad (3\text{-}3)$$

where p and q are constants.

$$1 = p([B]_0 - x) + q([A]_0 - x) \quad (3\text{-}4)$$

$$1 = p[B]_0 + q[A]_0 - x(p + q) \quad (3\text{-}5)$$

By setting $q = -p$, equation 3-5 can be reduced to $1 = p([B]_0 - [A]_0)$ and the following expressions are obtained for p and q:

$$p = \frac{1}{([B]_0 - [A]_0)} \quad (3\text{-}6)$$

$$q = -\frac{1}{([B]_0 - [A]_0)} \quad (3\text{-}7)$$

Substitution into equation 3-3 yields

$$\frac{1}{([A]_0 - x)([B]_0 - x)} = \frac{1}{([B]_0 - [A]_0)([A]_0 - x)} - \frac{1}{([B]_0 - [A]_0)([B]_0 - x)} \quad (3\text{-}8)$$

Equation 3-8 represents the left-hand side of equation 3-2 in a form which can be integrated:

$$\frac{1}{([B]_0 - [A]_0)} \int \frac{dx}{([A]_0 - x)} - \frac{1}{([B]_0 - [A]_0)} \int \frac{dx}{([B]_0 - x)} = k_1 \int dt \quad (3\text{-}9)$$

$$\frac{1}{([B]_0 - [A]_0)} \left[\int_0^x \frac{d([B]_0 - x)}{([B]_0 - x)} - \int_0^x \frac{d([A]_0 - x)}{([A]_0 - x)} \right] = k_1 \int_0^t dt \quad (3\text{-}10)$$

or

$$k_1 = \frac{1}{t} \frac{1}{([B]_0 - [A]_0)} \ln \left[\frac{([B]_0 - x)}{([A]_0 - x)} \right] + I \quad (3\text{-}11)$$

16 Second-order Reactions

The integration constant I is evaluated from the limiting condition that at $t = 0$, $x = 0$. Hence (3-11) reduces to

$$0 = \frac{1}{([B]_0 - [A]_0)} \ln \left[\frac{[B]_0}{[A]_0}\right] + I \qquad (3\text{-}12)$$

Substitution of the expression for I into (3-11) yields

$$k_1 t = \frac{1}{([B]_0 - [A]_0)} \left[\ln \frac{([B]_0 - x)}{([A]_0 - x)} - \ln \frac{[B]_0}{[A]_0}\right] \qquad (3\text{-}13)$$

$$k_1 t = \frac{1}{([B]_0 - [A]_0)} \ln \left[\frac{[A]_0([B]_0 - x)}{[B]_0([A]_0 - x)}\right] \qquad (3\text{-}14)$$

Inspection of this equation will make it apparent that it cannot be applied to systems in which the initial concentrations of both reactants are the same or in which only one reactant undergoes a second-order reaction. This is because when $[A]_0 = [B]_0$ the equation reduces to $k_1 t = 0$, which is indeterminate. To overcome this possibility, suitable conditions have to be imposed upon equation 3-1 before integration. These conditions are discussed in the next section.

SYSTEM 4. TYPE II

When two different species A and B at the same initial concentration undergo a second-order reaction, their concentrations remain equal throughout the reaction. The rate of reaction at any instant is then proportional to the square of the concentration of each reactant.

If A_0 represents the initial number of moles of the reacting species A and B, and x the number of moles of each that react in a time interval t, then the rate of formation of product can be described by the following mechanism and rate equation:

$$A + B \xrightarrow{k_2} C \qquad \text{(II)}$$

Amount at time $t = 0$ $[A]_0$ $[A]_0$ 0

Amount at time $t = t$ $([A]_0 - x)$ $([A]_0 - x)$ x

$$\frac{dx}{dt} = k_2([A]_0 - x)^2 \qquad (4\text{-}1)$$

$$\frac{dx}{([A]_0 - x)^2} = k_2 \, dt \qquad (4\text{-}2)$$

or

$$-\int \frac{d([A]_0 - x)}{([A]_0 - x)^2} = k_2 \int dt \qquad (4\text{-}3)$$

$$\frac{1}{([A]_0 - x)} + I = k_2 t \qquad (4\text{-}4)$$

The integration constant I is evaluated from the limiting condition that at $t = 0$, $x = 0$. Equation 4-4 then reduces to $I = -1/[A]_0$. Substituting for I back into (4-4) yields

$$\frac{1}{([A]_0 - x)} - \frac{1}{[A]_0} = \frac{x}{[A]_0([A]_0 - x)} = k_2 t \qquad (4\text{-}5)$$

or

$$k_2 = \frac{1}{t} \frac{x}{[A]_0([A]_0 - x)} \qquad (4\text{-}6)$$

If x is replaced by $([A]_0 - [A])$, equation 4-6 takes on the form

$$[A] = \left[\frac{[A]_0}{1 + k_2[A]_0 t} \right] \qquad (4\text{-}7)$$

SYSTEM 5. TYPE III

Another special type of second-order reactions includes systems involving a single reactant. In these reactions the rate at any instant is proportional to the square of the concentration of A. The derivation of the rate equation is identical with the one discussed previously, except that by convention the rate constant is usually written as $2k$, since two identical molecules disappear per collision.

$$A + A \xrightarrow{k_3} B \qquad \text{(III)}$$

Amount at $t = 0$ $\quad [A]_0 \qquad [A]_0 \qquad 0$
Amount at $t = t$ $\quad ([A]_0 - x) \quad ([A]_0 - x) \quad x$

$$\frac{dx}{dt} = 2k_3([A]_0 - x)^2 \qquad (5\text{-}1)$$

Hence

$$\int_{[A]_0}^{([A]_0 - x)} \frac{dx}{([A]_0 - x)^2} = 2k_3 \int_0^t dt \qquad (5\text{-}2)$$

$$\frac{1}{([A]_0 - x)} - \frac{1}{[A]_0} = 2k_3 t \qquad (5\text{-}3)$$

The half-life of the reaction is given by $t_{1/2} = 1/(2k_3[A]_0)$.

REFERENCES

1. N. Menschutkin, *Z. Phys. Chem.*, **6**, 41 (1890).
2. J. Walker and R. J. Hambly, *Trans. Chem. Soc.* (*London*), **67**, 746 (1895).
3. R. G. W. Norrish and F. P. Smith, *Trans. Chem. Soc.* (*London*), **129**, (1928).
4. J. C. Warner and F. B. Stitt, *J. Am. Chem. Soc.*, **55**, 4807 (1933).
5. H. Kaufmann and A. Wassermann, *Trans. Chem. Soc.* (*London*), **870**, (1939).
6. G. Czapski and B. H. J. Bielski, *J. Phys. Chem.*, **67**, 2180 (1963).
7. B. H. J. Bielski and A. O. Allen, *Proc. 2nd Tihany Symp. Radiation Chemistry*, Akademiai Biado, Budapest, 1967, p. 81.
8. B. H. J. Bielski and H. A. Schwarz, *J. Phys. Chem.*, **72**, 3836 (1968).
9. J. Rabani and S. O. Nielsen, *J. Phys. Chem.*, **73**, 3736 (1969).
10. D. Behar, G. Czapski, L. M. Dorfman, J. Rabani, and H. A. Schwarz, *J. Phys. Chem.*, **74**, 3209 (1970).
11. C. Capellos and A. O. Allen, *J. Phys. Chem.*, **74**, 840 (1970).
12. B. H. J. Bielski, D. A. Comstock, and R. A. Bowen, *J. Am. Chem. Soc.* **93**, 5624 (1971).

6

Third-order Reactions

A chemical reaction is said to be of third-order if its rate depends on three concentration terms. In solutions such reactions are so rare that only a few examples can be found in the literature.

Three types of third-order reactions are possible.

SYSTEM 6. TYPE I

In these reactions the initial concentrations of the reactants are equal:

$$A + B + C \xrightarrow{k_1} \text{product} \quad (I)$$

Amount at
 time $t = 0$ $[A]_0$ $[A]_0$ $[A]_0$ 0
Amount at
 time $t = t$ $([A]_0 - x)$ $([A]_0 - x)$ $([A]_0 - x)$ x

20 Third-order Reactions

The rate is

$$\frac{dx}{dt} = k_1([A]_0 - x)^3 \tag{6-1}$$

$$\int_0^x \frac{dx}{([A]_0 - x)^3} = k_1 \int_0^t dt \tag{6-2}$$

$$-\int_0^x \frac{d([A]_0 - x)}{([A]_0 - x)^3} = k_1 \int_0^t dt \tag{6-3}$$

$$\frac{1}{2([A]_0 - x)^2} + I = k_1 t \tag{6-4}$$

The integration constant I is evaluated from the boundary conditions; when $t = 0$, $x = 0$, and $I = -(1/2[A]_0^2)$. Substitution of this value into equation 6-4 yields

$$\frac{1}{2([A]_0 - x)^2} - \frac{1}{2[A]_0^2} = k_1 t \tag{6-5}$$

or

$$k_1 = \frac{1}{2t}\left[\frac{1}{([A]_0 - x)^2} - \frac{1}{[A]_0^2}\right] \tag{6-6}$$

and

$$t_{1/2} = \frac{3}{2k_1[A]_0^2} \tag{6-7}$$

SYSTEM 7. TYPE II

When two out of the three reactants are initially at equal concentrations, the third-order reaction process is formulated as follows:

$$A + A + B \xrightarrow{k_2} \text{product} \quad \text{(II)}$$

Amount at $t = 0$ $[A]_0$ $[B]_0$ 0
Amount at $t = t$ $([A]_0 - 2x)$ $([B]_0 - x)$ x

$$\frac{dx}{dt} = k_2([A]_0 - 2x)^2([B]_0 - x) \tag{7-1}$$

$$\frac{dx}{([A]_0 - 2x)^2([B]_0 - x)} = k_2\, dt \tag{7-2}$$

System 7. Type II

Equation 7-2 can be integrated after resolution into partial fractions:

$$\frac{1}{([A]_0 - 2x)^2([B]_0 - x)} = \frac{px + q}{([A]_0 - 2x)^2} + \frac{z}{([B]_0 - x)} \quad (7\text{-}3)$$

$$1 = px([B]_0 - x) + q([B]_0 - x) + z([A]_0 - 2x)^2 \quad (7\text{-}4)$$

This reduces to

$$1 = q[B]_0 + z[A]_0^2 + (p[B]_0 - q - 4z[A]_0)x + (4z - p)x^2 \quad (7\text{-}5)$$

from which p, q, and z can be evaluated:

$$q[B]_0 + [A]_0^2 z = 1 \quad (7\text{-}6)$$

hence

$$q = \frac{1 - [A]_0^2 z}{[B]_0} \quad (7\text{-}7)$$

Since

$$4z - p = 0 \quad \text{or} \quad p = 4z \quad (7\text{-}8)$$

hence

$$p[B]_0 - q - 4z[A]_0 = 0 \quad (7\text{-}9)$$

Substitution of the values for p and q into the last equation yields z in terms of $[A]_0$ and $[B]_0$:

$$4z[B]_0 - \frac{(1 - [A]_0^2 z)}{[B]_0} - 4z[A]_0[B]_0 = 0 \quad (7\text{-}10)$$

or

$$z = \frac{1}{(2[B]_0 - [A]_0)^2} \quad (7\text{-}11)$$

To summarize:

$$p = \frac{4}{(2[B]_0 - [A]_0)^2} \quad (7\text{-}12)$$

$$q = \frac{4([B]_0 - [A]_0)}{(2[B]_0 - [A]_0)^2} \quad (7\text{-}13)$$

$$z = \frac{1}{(2[B]_0 - [A]_0)^2} \quad (7\text{-}14)$$

22   Third-order Reactions

and equation 7-2 in terms of the partial fractions becomes

$$\int_0^x \frac{(px+q)\,dx}{([A]_0-2x)^2} + \int_0^x \frac{z\,dx}{([B]_0-x)} = k_2\int_0^t dt \quad (7\text{-}15)$$

Substituting for p, q, and z equation 7-15 becomes

$$\int_0^x \frac{\{4x/(2[B]_0-[A]_0)^2\} + \{4([B]_0-[A]_0)/(2[B]_0-[A]_0)^2\}}{([A]_0-2x)^2}\,dx$$

$$+\int_0^x \frac{\{1/(2[B]_0-[A]_0)^2\}}{([B]_0-x)}\,dx = k_2\int_0^t dt \quad (7\text{-}16)$$

$$\int_0^x \frac{4x\,dx}{(2[B]_0-[A]_0)^2([A]_0-2x)^2} + \int_0^x \frac{4([B]_0-[A]_0)\,dx}{(2[B]_0-[A]_0)^2([A]_0-2x)^2}$$

$$+\int_0^x \frac{dx}{(2[B]_0-[A]_0)^2([B]_0-x)} = k_2\int_0^t dt \quad (7\text{-}17)$$

$$-\int_0^x \frac{2x\,d([A]_0-2x)}{([A]_0-2x)^2} - ([B]_0-[A]_0)\int_0^x \frac{2\,d([A]_0-2x)}{([A]_0-2x)^2}$$

$$-\int_0^x \frac{d([B]_0-x)}{([B]_0-x)} = (2[B]_0-[A]_0)^2 k_2\int_0^t dt \quad (7\text{-}18)$$

For convenience the three integrals on the left-hand side of equation 7-18 are solved separately and referred to as $(7\text{-}18)_1$, $(7\text{-}18)_2$, and $(7\text{-}18)_3$, counting from left to right. The first integral is solved by setting $(A_0 - 2x) = w$, so that

$$(7\text{-}18)_1 = -\int \frac{2x\,d([A]_0-2x)}{([A]_0-2x)^2} = -\int \frac{([A]_0-w)\,dw}{(w)^2}$$

$$= -\left[\int \frac{[A]_0\,dw}{(w)^2} - \int \frac{dw}{(w)}\right]$$

$$= \frac{[A]_0}{w}\Big|_{w_1=[A]_0}^{w_2=([A]_0-2x)} + \ln(w)\Big|_{w_1=[A]_0}^{w_2=([A]_0-2x)}$$

$$= [A]_0\left[\frac{1}{([A]_0-2x)} - \frac{1}{[A]_0}\right] + \ln([A]_0-2x) - \ln[A]_0$$

$$= \ln\left[\frac{([A]_0-2x)}{[A]_0}\right] + \left[\frac{2[A]_0 x}{[A]_0([A]_0-2x)}\right]$$

The solution for the second (7-18)$_2$ and third (7-18)$_3$ integrals are simple:

$$(7\text{-}18)_2 = -2([B]_0 - [A]_0) \int_0^x \frac{d([A]_0 - 2x)}{([A]_0 - 2x)^2}$$

$$= \frac{2([B]_0 - [A]_0)}{([A]_0 - 2x)} \Big|_0^x$$

$$= 2([B]_0 - [A]_0)\left[\frac{1}{([A]_0 - 2x)} - \frac{1}{[A]_0}\right]$$

$$= \frac{2([B]_0 - [A]_0)2x}{[A]_0([A]_0 - 2x)}$$

$$(7\text{-}18)_3 = -\int_0^x \frac{d([B]_0 - x)}{([B]_0 - x)} = -\ln([B]_0 - x) - \ln[B]_0$$

$$= \ln\left[\frac{[B]_0}{([B]_0 - x)}\right]$$

Substitution of the solutions back into equation 7-18 yields

$$\ln\frac{([A]_0 - 2x)}{[A]_0} + \frac{2[A]_0 x}{[A]_0([A]_0 - 2x)} + \frac{2x(2[B]_0 - 2[A]_0)}{[A]_0([A]_0 - 2x)}$$

$$+ \ln\frac{[B]_0}{([B]_0 - x)} = (2[B]_0 - [A]_0)^2 k_2 t \quad (7\text{-}19)$$

Hence

$$k_2 = \frac{1}{t(2[B]_0 - [A]_0)^2}\left[\frac{2x(2[B]_0 - [A]_0)}{[A]_0([A]_0 - 2x)} + \ln\frac{[B]_0([A]_0 - 2x)}{[A]_0([B]_0 - x)}\right]$$

$$(7\text{-}20)$$

SYSTEM 8. TYPE III

This is the case when all three reactants have different initial concentrations. The mathematical description of such a mechanism is very similar to the previous case:

24 Third-order Reactions

$$A + B + C \xrightarrow{k_3} \text{product} \quad \text{(III)}$$

Amount at
time $t = 0$ $[A]_0$ $[B]_0$ $[C]_0$ 0

Amount at
time $t = t$ $([A]_0 - x)$ $([B]_0 - x)$ $([C]_0 - x)$ x

Hence

$$\frac{dx}{([A]_0 - x)([B]_0 - x)([C]_0 - x)} = k_3 \, dt \quad (8\text{-}1)$$

To integrate equation 8-1, its left-hand side has to be resolved into partial fractions:

$$\frac{1}{([A]_0 - x)([B]_0 - x)([C]_0 - x)} = \frac{p}{([A]_0 - x)} + \frac{q}{([B]_0 - x)}$$

$$+ \frac{z}{([C]_0 - x)} \quad (8\text{-}2)$$

hence

$$1 = p([B]_0 - x)([C]_0 - x) + q([A]_0 - x)([C]_0 - x)$$
$$+ z([A]_0 - x)([B]_0 - x) \quad (8\text{-}3)$$

$$1 = p[B]_0[C]_0 + q[A]_0[C]_0 + z[A]_0[B]_0$$
$$- x(p[B]_0 + p[C]_0 + q[A]_0 + q[C]_0 + z[A]_0 + z[B]_0)$$
$$+ x^2(p + q + z) \quad (8\text{-}4)$$

To express p, q, and z in terms of $[A]_0$, $[B]_0$, and $[C]_0$ the following mathematical operations are carried out:

1. Set

$$p[B]_0[C]_0 + q[A]_0[C]_0 + z[A]_0[B]_0 = 1 \quad (8\text{-}5)$$

$$p([B]_0 + [C]_0) + q([A]_0 + [C]_0) + z([A]_0 + [B]_0) = 0 \quad (8\text{-}6)$$

$$p + q + z = 0 \quad (8\text{-}7)$$

2. From these equations

$$p = \left[\frac{1 - q[A]_0[C]_0 - z[A]_0[B]_0}{[B]_0[C]_0}\right] \quad (8\text{-}8)$$

$$= \left[\frac{1 - [A]_0(q[C]_0 + z[B]_0)}{[B]_0[C]_0}\right] \quad (8\text{-}9)$$

and

$$p = -(q + z) \quad (8\text{-}10)$$

$$p([B]_0 + [C]_0) + [A]_0(q + z) + q[C]_0 + z[B]_0 = 0 \quad (8\text{-}11)$$

$$p([B]_0 + [C]_0) - p[A]_0 + q[C]_0 + z[B]_0 = 0 \quad (8\text{-}12)$$

$$p([B]_0 + [C]_0) - p[A]_0 + \frac{(1 - p[B]_0[C]_0)}{[A]_0} = 0 \quad (8\text{-}13)$$

Hence

$$p = -\frac{1}{([A]_0 - [B]_0)([C]_0 - [A]_0)}$$

$$= -\frac{([B]_0 - [C]_0)}{([A]_0 - [B]_0)([B]_0 - [C]_0)([C]_0 - [A]_0)} \quad (8\text{-}14)$$

3. By the use of similar substitutions

$$q = -\frac{1}{([A]_0 - [B]_0)([B]_0 - [C]_0)}$$

$$= -\frac{([C]_0 - [A]_0)}{([A]_0 - [B]_0)([B]_0 - [C]_0)([C]_0 - [A]_0)} \quad (8\text{-}15)$$

and

$$z = -\frac{1}{([B]_0 - [C]_0)([C]_0 - [A]_0)}$$

$$= -\frac{([A]_0 - [B]_0)}{([A]_0 - [B]_0)([B]_0 - [C]_0)([C]_0 - [A]_0)} \quad (8\text{-}16)$$

Substitution of the expressions found for p, q, and z into (8-2)

26 Third-order Reactions

and rearrangement yields a differential equation which is equivalent to equation 8-1:

$$-\frac{([B]_0 - [C]_0)}{([A]_0 - [B]_0)([B]_0 - [C]_0)([C]_0 - [A]_0)} \frac{dx}{([A]_0 - x)}$$

$$-\frac{([C]_0 - [A]_0)}{([A]_0 - [B]_0)([B]_0 - [C]_0)([C]_0 - [A]_0)} \frac{dx}{([B]_0 - x)}$$

$$-\frac{([A]_0 - [B]_0)}{([A]_0 - [B]_0)([B]_0 - [C]_0)([C]_0 - [A]_0)} \frac{dx}{([C]_0 - x)} = k_3 \, dt$$

(8-17)

$$\frac{([B]_0 - [C]_0)}{([A]_0 - [B]_0)([B]_0 - [C]_0)([C]_0 - [A]_0)} \int_0^x \frac{d([A]_0 - x)}{([A]_0 - x)}$$

$$+ \frac{([C]_0 - [A]_0)}{([A]_0 - [B]_0)([B]_0 - [C]_0)([C]_0 - [A]_0)} \int_0^x \frac{d([B]_0 - x)}{([B]_0 - x)}$$

$$+ \frac{[A]_0 - [B]_0}{([A]_0 - [B]_0)([B]_0 - [C]_0)([C]_0 - [A]_0)} \int_0^x \frac{d([C]_0 - x)}{([C]_0 - x)} = k_3 \int_0^x dt$$

(8-18)

Hence

$$\frac{([B]_0 - [C]_0)}{([A]_0 - [B]_0)([B]_0 - [C]_0)([C]_0 - [A]_0)} \ln \left[\frac{([A]_0 - x)}{[A]_0}\right]$$

$$+ \frac{([C]_0 - [A]_0)}{([A]_0 - [B]_0)([B]_0 - [C]_0)([C]_0 - [A]_0)} \ln \left[\frac{([B]_0 - x)}{[B]_0}\right]$$

$$+ \frac{([A]_0 - [B]_0)}{([A]_0 - [B]_0)([B]_0 - [C]_0)([C]_0 - [A]_0])} \ln \left[\frac{[C]_0 - x)}{[C]_0}\right] = k_3 t$$

(8-19)

or

$$k_3 = \frac{([B]_0 - [C]_0) \ln \{[([A]_0 - x)/[A]_0]\} + ([C]_0 - [A]_0) \ln \{[([B]_0 - x)/[B]_0]\} + ([A]_0 - [B]_0) \ln \{[([C]_0 - x)/[C]_0]\}}{t([A]_0 - [B]_0)([B]_0 - [C]_0)([C]_0 - [A]_0)}$$

(8-20)

REFERENCES

1. H. A. Liebhafsky and A. Mohammad, *J. Am. Chem. Soc.*, **55,** 3977 (1933).
2. H. A. Liebhafsky and A. Mohammad, *J. Phys. Chem.*, **38,** 857 (1934).

7

Reaction of *n*-th Order

SYSTEM 9

Although reactions of higher orders than the third are unknown, it is important to analyze reactions of arbitrary order *n* since fractional orders are not uncommon.

For a reaction of *n*th order involving a single reactant

$$nA \xrightarrow{k_1} \text{products} \qquad (I)$$

the rate of disappearance of A is given by

$$-\frac{d[A]}{dt} = k_1[A]^n \qquad (9\text{-}1)$$

Equation 9-1 after rearrangement can be integrated

$$\frac{d[A]}{[A]^n} = -k_1\, dt \qquad (9\text{-}2)$$

$$\int_{[A]_0}^{[A]} [A]^{-n}\, d[A] = -k_1 \int_0^t dt \qquad (9\text{-}3)$$

$$\left. \frac{[A]^{-n+1}}{-n+1} \right|_{[A]_0}^{[A]} = k_1 t \bigg|_0^t \qquad (9\text{-}4)$$

$$[A]^{-(n-1)} - [A]_0^{-(n-1)} = (n-1)k_1 t \qquad (9\text{-}5)$$

$$\frac{1}{[A]^{n-1}} - \frac{1}{[A]_0^{n-1}} = (n-1)k_1 t \qquad (9\text{-}6)$$

The corresponding half-life time $t_{1/2}$, during which initial concentration is halved, is computed by substituting $[A]_0/2$ for $[A]_0$ into equation 9-6:

$$\frac{1}{[([A]_0/2)]^{n-1}} - \frac{1}{[A]_0^{n-1}} = (n-1)k_1 t_{1/2} \qquad (9\text{-}7)$$

$$\frac{2^{n-1}}{[A]_0^{n-1}} - \frac{1}{[A]_0^{n-1}} = \frac{2^{n-1}-1}{[A]_0^{n-1}} = (n-1)k t_{1/2} \qquad (9\text{-}8)$$

hence

$$t_{1/2} = \left[\frac{2^{(n-1)}-1}{(n-1)k_1[A]_0^{n-1}}\right] \qquad (9\text{-}9)$$

This derivation is valid only for $n \neq 1$.

8

Reversible Reactions

Reversible or opposing reactions are those in which the products can be reconverted back to the original reactants. In a system of reversible reactions the apparent rate of the forward reaction will decrease as the reaction products accumulate until eventually a state of dynamic equilibrium is established. At equilibrium the forward and backward reactions proceed at equal rates.

SYSTEM 10. FIRST-ORDER REVERSIBLE REACTION

$$A \underset{k_2}{\overset{k_1}{\rightleftharpoons}} B \qquad (I)$$

This mechanism describes the simplest case of opposing reactions which is that of two first-order reactions. The true rate of concentration change for either reactant is the difference in the rates in the opposite directions, each proportional to the concentration of the reacting compound. The differential equations for this mechanism are

$$-\frac{d[A]}{dt} = k_1[A] - k_2[B] \qquad (10\text{-}1)$$

$$-\frac{d[B]}{dt} = k_2[B] - k_1[A] \qquad (10\text{-}2)$$

System 10. First-order Reversible Reaction

Mathematical solutions of these equations can be obtained either by the classic method or by application of Laplace transforms. Since the mathematical operations differ with the initial boundary conditions imposed upon the mechanism, the following situations are discussed:

Case I.   Initially only one substance is present in the system.
Case II. Reactants A and B are both present at the commencement of the reaction.
Case III. System approaches equilibrium.

Case I

This case deals with the situation where initially only A is present, that is, at time $t = 0$, the concentration of A is $[A]_0$ and $[B]_0 = 0$.

CLASSICAL SOLUTION.   From the material balance equation

$$[A]_0 - [A] - [B] = 0 \tag{10-3}$$

the expression $[A]_0 - [A] = [B]$ is substituted into (10-1) to give

$$-\frac{d[A]}{dt} = (k_1 + k_2)[A] - k_2[A]_0 \tag{10-4}$$

which can be integrated directly:

$$-\int \frac{d[A]}{(k_1 + k_2)[A] - k_2[A]_0} = \int dt \tag{10-5}$$

$$\frac{1}{(k_1 + k_2)} \int_{[A]_0}^{[A]} \frac{d[(k_1 + k_2)[A] - k_2[A]_0]}{[(k_1 + k_2)[A] - k_2[A]_0]} = -\int_0^t dt \tag{10-6}$$

$$\ln[(k_1 + k_2)[A] - k_2[A]_0] - \ln[(k_1 + k_2)[A]_0 - k_2[A]_0]$$
$$= -(k_1 + k_2)t \tag{10-7}$$

$$\ln\left[\frac{(k_1 + k_2)[A] - k_2[A]_0}{k_1[A]_0}\right] = -(k_1 + k_2)t \tag{10-8}$$

32   Reversible Reactions

The exponential form of (10-8) is

$$[A] = \frac{[A]_0}{(k_1 + k_2)} [k_2 + k_1 e^{-(k_1+k_2)t}] \quad (10\text{-}9)$$

Equation 10-8 can also be expressed in terms of the equilibrium constant $K = k_1/k_2$. In dividing the left-hand side (numerator and denominator) by k_2, and substituting $-[B] = [A] - [A]_0$, gives

$$\ln\left[\frac{[A]K - [B]}{[A]_0 K}\right] = -(k_1 + k_2)t \quad (10\text{-}10)$$

An equation similar to (10-8) can be derived for $[B]$ if $[A]_0 - [B]$ is substituted for $[A]$:

$$[A]_0 - [B] = \frac{[A]_0}{(k_1 + k_2)} [k_2 + k_1 e^{-(k_1+k_2)t}] \quad (10\text{-}11)$$

$$[B] = \frac{k_1[A]_0}{(k_1 + k_2)} [1 - e^{-(k_1+k_2)t}] \quad (10\text{-}12)$$

OPERATOR METHOD. This derivation is based on the assumption, that at time $t = 0$, $[B]_0 = [B] = 0$, and that the rates at which A is converted into B and B into A are directly proportional to the concentrations of the respective substances. Hence equation 10-1 and 10-2 can be transformed (see p. 50) to give

$$S[A] - S[A]_0 = k_2[B] - k_1[A] \quad (10\text{-}13)$$

and

$$S[B] = k_1[A] - k_2[B] \quad (10\text{-}14)$$

$$[B] = \frac{k_1[A]}{(S + k_2)} \quad (10\text{-}15)$$

Substituting $[B]$ into (10-13) yields, after rearrangement,

$$[A] = \frac{[A]_0(S + k_2)}{(S + k_1 + k_2)} \quad (10\text{-}16)$$

System 10. First-order Reversible Reaction

Similarly, substitution of the expression for [A] into (10-15) leads to an expression for [B]:

$$[B] = \frac{k_1[A]_0}{(S + k_1 + k_2)} \qquad (10\text{-}17)$$

Replacing the transforms in equations 10-16 and 10-17 by the originals listed in Appendix II, yields equations 10-9 and 10-12.

Case II

If the initial conditions are such that at time $t = 0$, the concentration of A is $[A]_0$ and the concentration of B is $[B]_0$ then at any time t the following relationship holds true:

$$[A]_0 - [A] = [B] - [B]_0 \qquad (10\text{-}18)$$

or

$$[B] = [B]_0 + [A]_0 - [A] \qquad (10\text{-}19)$$

Substituting [B] into equation (10-1) yields

$$-\frac{d[A]}{dt} = k_1[A] - k_2[[B]_0 + [A]_0 - [A]] \qquad (10\text{-}20)$$

or

$$\frac{d[A]}{dt} = k_2[[A]_0 + [B]_0] - (k_1 + k_2)[A] \qquad (10\text{-}21)$$

Upon rearrangement, (10-1) can be integrated directly

$$\int_{[A]_0}^{[A]} \frac{d[A]}{k_2([A]_0 + [B]_0) - (k_1 + k_2)[A]} = \int_0^t dt \qquad (10\text{-}22)$$

$$-\frac{1}{(k_1 + k_2)} \int_{[A]_0}^{[A]} \frac{d[k_2([A]_0 + [B]_0) - (k_1 + k_2)[A]]}{k_2([A]_0 + [B]_0) - (k_1 + k_2)[A]} = \int_0^t dt \qquad (10\text{-}23)$$

$$\ln\left[\frac{k_2([A]_0 + [B]_0 - [A]) - k_1[A]}{k_2[B]_0 - k_1[A]_0}\right] = -(k_1 + k_2)t \qquad (10\text{-}24)$$

and finally

$$\ln\left[\frac{k_1[A] - k_2[B]}{k_1[A]_0 - k_2[B]_0}\right] = -(k_1 + k_2)t \qquad (10\text{-}25)$$

34 Reversible Reactions

Equation 10-25 can also be expressed in terms of the equilibrium constant $K = k_1/k_2$, by dividing the left-hand side (numerator and denominator) by k_2,

$$\ln\left[\frac{[A]K - [B]}{[A]_0 K - [B]_0}\right] = -(k_1 + k_2)t \qquad (10\text{-}26)$$

Case III

An alternate analysis of the rates of first-order opposing reactions employs the symbol x, defined as the number of moles per liter of A converted to B at time t. At time $t = 0$, the concentrations of the two substances are $[A]_0$ and $[B]_0$ as in previous cases. Hence in terms of x, the mechanism can be described by the following differential equation:

$$\frac{dx}{dt} = k_1([A]_0 - x) - k_2([B]_0 + x) \qquad (10\text{-}27)$$

$$\frac{dx}{dt} = k_1[A]_0 - k_2[B]_0 - (k_1 + k_2)x \qquad (10\text{-}28)$$

After rearrangement, equation 10-28 can be integrated directly:

$$\int_0^x \frac{dx}{k_1[A]_0 - k_2[B]_0 - (k_1 + k_2)x} = \int_0^t dt \qquad (10\text{-}29)$$

$$-\frac{1}{(k_1 + k_2)} \int_0^x \frac{d[k_1[A]_0 - k_2[B]_0 - (k_1 + k_2)x]}{[k_1[A]_0 - k_2[B]_0 - (k_1 + k_2)x]} = \int_0^t dt \qquad (10\text{-}30)$$

$$\ln\left[\frac{k_1[A]_0 - k_2[B]_0}{k_1[A]_0 - k_2[B]_0 - (k_1 + k_2)x}\right] = (k_1 + k_2)t \qquad (10\text{-}31)$$

As time increases indefinitely, the reaction comes to a state of equilibrium and x attains a value x_e which can be considered a constant if the experimental conditions are kept constant. Furthermore, under these conditions $dx/dt = 0$ and the rates of the two opposing reactions are equal, and equation 10-27 yields

$$k_1([A]_0 - x_e) = k_2([B]_0 + x_e) \qquad (10\text{-}32)$$

The equilibrium constant is given by

$$K = \frac{k_1}{k_2} = \left[\frac{[B]_0 + x_e}{[A]_0 - x_e}\right] \quad (10\text{-}33)$$

Substitution of (10-32) into (10-31) yields

$$\ln\left[\frac{x_e}{x_e - x}\right] = (k_1 + k_2)t \quad (10\text{-}34)$$

from which the sum of the rate constants ($k_1 + k_2$) is determined.

In practice a plot of log ($x_e - x$) against time gives a straight line with a slope equal to $-0.4343(k_1 + k_2)$. The individual rate constants k_1 and k_2 can be determined if in addition to this result the ratio k_1/k_2 is known from measurements of equilibrium concentrations.

REFERENCES

1. P. Henry, *Z. Phys. Chem.*, **10**, 96, 98 (1892).
2. T. M. Lowry, *J. Chem. Soc.*, **75**, 211 (1899).
3. T. M. Lowry, *Trans. Chem. Soc. (London)*, **75**, 212 (1899).
4. J. C. Kendrew and E. A. Moelwyn-Hughes, *Proc. Roy. Soc. (London)*, **A176**, 352 (1940).
5. F. A. Long, F. B. Dunkle, and W. F. McDevit, *J. Phys. Colloid Chem.* **55**, 814, 829 (1951).
6. K. Micka, *Chem. Listy*, **48**, 355 (1954).

SYSTEM 11. FIRST-ORDER TWO-STAGE REVERSIBLE REACTION

$$A \underset{k_2}{\overset{k_1}{\rightleftarrows}} B \underset{k_4}{\overset{k_3}{\rightleftarrows}} C \quad (I)$$

The set of differential equations 11-1, 11-2, 11-3, which describe the change of concentration of *A*, *B*, and *C* with time,

Reversible Reactions

can be solved by the operator method.

$$\frac{d[A]}{dt} = k_2[B] - k_1[A] \tag{11-1}$$

$$\frac{d[B]}{dt} = k_1[A] - k_2[B] - k_3[B] + k_4[C] \tag{11-2}$$

$$\frac{d[C]}{dt} = k_3[B] - k_4[C] \tag{11-3}$$

Assuming that at time $t = 0$, $[A] = [A]_0$, and $[B]_0 = [C]_0 = 0$, and that the amounts of A, B, and C, which have reacted in time $t = t$, are $\{[A]_0 - [A]\}$, $\{([A]_0 - [A]) - [B]\}$, and $\{([A]_0 - [A] - [B] - [C]\} = 0$ respectively, the transformed functions for (11-1), (11-2), and (11-3) are

$$S[A] - S[A]_0 = k_2[B] - k_1[A] \tag{11-4}$$

$$S[B] = k_1[A] - k_2[B] - k_3[B] + k_4[C] \tag{11-5}$$

$$S[C] = k_3[B] - k_4[C] \tag{11-6}$$

(For definition of S see Appendix II)

Solving these equations in succession yields

$$[A] = \frac{[A]_0[S^2 + S(k_2 + k_3 + k_4) + k_2 k_4]}{[S^2 + S(k_1 + k_2 + k_3 + k_4) + k_1 k_3 + k_2 k_4 + k_1 k_4]} \tag{11-7}$$

$$[B] = \frac{[A]_0 k_1 (S + k_4)}{[S^2 + S(k_1 + k_2 + k_3 + k_4) + k_1 k_3 + k_2 k_4 + k_1 k_4]} \tag{11-8}$$

$$[C] = \frac{[A]_0 k_1 k_3}{[S^2 + S(k_1 + k_2 + k_3 + k_4) + k_1 k_3 + k_2 k_4 + k_1 k_4]} \tag{11-9}$$

System 11. First-order Two-stage Reversible Reaction

As is apparent, equations 11-7, 11-8, and 11-9 have the same denominator. This can be converted by factoring to $(S + \gamma_1)(S + \gamma_2)$, if one recognizes γ_1 and γ_2 as the negative roots of the quadratic equation

$$\gamma^2 + \gamma(k_1 + k_2 + k_3 + k_4) + k_1k_3 + k_2k_4 + k_1k_4 = 0 \tag{11-10}$$

Hence

$$[A] = \frac{[A]_0[S^2 + S(k_2 + k_3 + k_4) + k_2k_4]}{(S + \gamma_1)(S + \gamma_2)} \tag{11-11}$$

$$[B] = \frac{[A]_0 k_1(S + k_4)}{(S + \gamma_1)(S + \gamma_2)} \tag{11-12}$$

$$[C] = \frac{[A]_0 k_1 k_3}{(S + \gamma_1)(S + \gamma_2)} \tag{11-13}$$

Replacing the transforms by the originals yields

$$[A] = [A]_0 \left[\frac{k_2 k_4}{\gamma_1 \gamma_2} + \left(\frac{\gamma_1^2 - \gamma_1(k_2 + k_3 + k_4) + k_2 k_4}{\gamma_1(\gamma_1 - \gamma_2)} \right) e^{-\gamma_1 t} \right.$$
$$\left. + \left(\frac{\gamma_2^2 - \gamma_2(k_2 + k_3 + k_4) + k_2 k_4}{\gamma_2(\gamma_2 - \gamma_1)} \right) e^{-\gamma_2 t} \right] \tag{11-14}$$

$$[B] = k_1[A]_0 \left[\frac{k_4}{\gamma_1 \gamma_2} + \left(\frac{k_4 - \gamma_1}{\gamma_1(\gamma_1 - \gamma_2)} \right) e^{-\gamma_1 t} \right.$$
$$\left. + \left(\frac{k_4 - \gamma_2}{\gamma_2(\gamma_2 - \gamma_1)} \right) e^{-\gamma_2 t} \right] \tag{11-15}$$

$$[C] = k_1 k_3 [A]_0 \left[\frac{1}{\gamma_1 \gamma_2} + \left(\frac{1}{\gamma_1(\gamma_1 - \gamma_2)} \right) e^{-\gamma_1 t} \right.$$
$$\left. + \left(\frac{1}{\gamma_2(\gamma_2 - \gamma_1)} \right) e^{-\gamma_2 t} \right] \tag{11-16}$$

REFERENCES

1. A. Rakowski, *Z. Phys. Chem.*, **57**, 321 (1906).
2. T. M. Lowry and W. T. John, *J. Chem. Soc.*, **97**, 2634 (1910).
3. A. Skrabal, *Z. Phys. Chem.*, **83**, 247 (1929).
4. A. Skrabal, *Z. Electrochem.*, **42**, 228 (1936).
5. A. Skrabal, *Z. Electrochem.*, **43**, 309 (1937).
6. G. N. Vriens, *Ind. Eng. Chem.*, **46**, 669 (1954).
7. F. Halla, *Monatsh.*, **33**, 1448 (1956).
8. J. M. Los, L. B. Simpson, and K. Wiesner, *J. Am. Chem. Soc.*, **78**, 1564 (1956).

SYSTEM 12. SECOND-ORDER REVERSIBLE REACTION

$$A + B \underset{k_2}{\overset{k_1}{\rightleftharpoons}} C + D \qquad (I)$$

At time $t = 0$ $\quad [A]_0 \quad\quad B_0 \quad\quad 0 \quad 0$
At time $t = t$ $\quad ([A]_0 - x) \quad ([B]_0 - x) \quad x \quad x$

If at the commencement of the reaction the concentration of $[A]_0 = [B]_0$, and the products are absent altogether, then the mechanism above can be described by the following differential equation:

$$\frac{dx}{dt} = k_1([A]_0 - x)^2 - k_2(x)^2 \qquad (12\text{-}1)$$

When the system reaches equilibrium, that is, $dx/dt = 0$, equation 12-1 reduces to

$$k_1([A]_0 - x_e)^2 = k_2(x_e)^2 \qquad (12\text{-}2)$$

or

$$k_2 = \frac{k_1([A]_0 - x_e)^2}{(x_e)^2} \qquad (12\text{-}3)$$

System 12. Second-order Reversible Reaction

where x_e is the equilibrium concentration of C and D. Substituting the expression for k_2 into equation 12-1 yields

$$\frac{dx}{dt} = k_1([A]_0 - x)^2 - \frac{k_1([A]_0 - x_e)^2(x)^2}{(x_e)^2} \qquad (12\text{-}4)$$

$$\int_0^x \frac{(x_e)^2\, dx}{([A]_0 - x)^2(x_e)^2 - ([A]_0 - x_e)^2(x)^2} = k_1 \int_0^t dt \qquad (12\text{-}5)$$

To integrate equation 12-5, its left-hand side has to be rearranged and resolved into partial fractions:

1. Rearrangement of the denominator:

$$\text{Denominator} = ([A]_0 - x)^2(x_e)^2 - ([A]_0 - x_e)^2(x)^2 \qquad (12\text{-}6)$$
$$= [A]_0^2 x_e^2 - 2[A]_0 x_e^2 x - [A]_0^2 x^2 + 2[A]_0 x_e x^2 \qquad (12\text{-}7)$$
$$= [A]_0^2(x_e^2 - x^2) - 2[A]_0 x_e x(x_e - x) \qquad (12\text{-}8)$$

Since $(x_e^2 - x^2) = (x_e + x)(x_e - x)$, the denominator can be written in the form;

$$\text{Denominator} = (x_e - x)[[A]_0^2 x_e + x([A]_0^2 - 2[A]_0 x_e)] \qquad (12\text{-}9)$$

2. Resolution into partial fractions:

$$\frac{(x_e)^2}{(x_e - x)[[A]_0^2 x_e + x([A]_0^2 - 2[A]_0 x_e)]} = \frac{p}{(x_e - x)}$$
$$+ \frac{q}{[A]_0^2 x_e + x([A]_0^2 - 2[A]_0 x_e)} \qquad (12\text{-}10)$$

$$(x_e)^2 = p([A]_0^2 x_e + [A]_0^2 x - 2[A]_0 x_e x) + q(x_e - x) \qquad (12\text{-}11)$$
$$(x_e)^2 = p[A]_0^2 x_e + qx_e + x(p[A]_0^2 - 2p[A]_0 x_e - q) \qquad (12\text{-}12)$$

In setting

$$p[A]_0^2 - 2p[A]_0 x_e - q = 0 \qquad (12\text{-}13)$$

one finds that

$$q = p[A]_0^2 - 2p[A]_0 x_e \qquad (12\text{-}14)$$

and

$$p[A]_0^2 x_e + qx_e = x_e^2 \qquad (12\text{-}15)$$

40 Reversible Reactions

Hence after substitution of the expression for q one obtains

$$p[A]_0^2 x_e + x_e(p[A]_0^2 - 2p[A]_0 x_e) = x_e^2 \qquad (12\text{-}16)$$

$$p = \frac{x_e}{2[A]_0([A]_0 - x_e)} \qquad (12\text{-}17)$$

$$q = \frac{x_e([A]_0 - 2x_e)}{2([A]_0 - x_e)} \qquad (12\text{-}18)$$

Substitution of the partial fractions and the expressions found for p and q into equation 12-5 yields

$$\frac{x_e}{2[A]_0([A]_0 - x_e)} \int_0^x \frac{dx}{(x_e - x)} + \frac{x_e([A]_0 - 2x_e)}{2[A]_0([A]_0 - x_e)}$$

$$\times \int_0^x \frac{dx}{[A]_0 x_e + x([A]_0 - 2x_e)} = k_1 \int_0^t dt \qquad (12\text{-}19)$$

While the first term on the left-hand side is integrated in the conventional manner, the solution for the second term can be found in standard integral tables:

$$\int \frac{du}{(a + bu)} = \frac{1}{b} \ln(a + bu) + \text{constant}$$

Hence

$$-\frac{x_e}{2[A]_0([A]_0 - x_e)} \ln\left[\frac{(x_e - x)}{x_e}\right] + \frac{x_e}{2[A]_0([A]_0 - x_e)}$$

$$\times \{\ln[[A]_0 x_e + x([A]_0 - 2x_e)] - \ln[A]_0 - \ln x_e\} = k_1 t \qquad (12\text{-}20)$$

or

$$k_1 = \frac{1}{t} \frac{x_e}{2[A]_0([A]_0 - x_e)} \ln\left[\frac{[A]_0 x_e + x([A]_0 - 2x_e)}{[A]_0(x_e - x)}\right] \qquad (12\text{-}21)$$

REFERENCE

1. S. W. Benson, *The Foundations of Chemical Kinetics*, McGraw-Hill, New York, 1960.

SYSTEM 13. FIRST- AND SECOND-ORDER REVERSIBLE REACTIONS

$$A \underset{k_2}{\overset{k_1}{\rightleftharpoons}} B + C \qquad (I)$$

At time $t = 0$ $\quad [A]_0 \quad\quad 0 \quad 0$
At time $t = t$ $\quad ([A]_0 - x) \quad x \quad x$

If it is assumed that B and C are absent at the commencement of the reaction, then the net rate of the forward reaction is described by the following equation:

$$\frac{dx}{dt} = k_1([A]_0 - x) - k_2 x^2 \qquad (13\text{-}1)$$

Since at equilibrium the forward and reverse reactions have equal velocities,

$$k_1([A]_0 - x_e) = k_2 x_e^2 \qquad (13\text{-}2)$$

where x_e represents the concentration of B and C at equilibrium. Hence

$$k_2 = \frac{k_1([A]_0 - x_e)}{x_e^2} \qquad (13\text{-}3)$$

and

$$\frac{dx}{dt} = k_1([A]_0 - x) - \frac{k_1([A]_0 - x_e)x^2}{x_e^2} \qquad (13\text{-}4)$$

or

$$\frac{x_e^2 \, dx}{([A]_0 - x)x_e^2 - ([A]_0 - x_e)x^2} = k_1 \, dt \qquad (13\text{-}5)$$

$$\int_0^x \frac{x_e^2 \, dx}{(x_e - x)([A]_0 x_e + [A]_0 x - x_e x)} = k_1 \int_0^t dt \qquad (13\text{-}6)$$

The left-hand side of equation 13-6 can be integrated after resolution into partial fractions:

$$\frac{x_e^2}{(x_e - x)([A]_0 x_e + [A]_0 x - x_e x)} = \frac{p}{(x_e - x)} + \frac{q}{([A]_0 x_e + [A]_0 x - x_e x)} \qquad (13\text{-}7)$$

$$x_e^2 = p(x_e + [A]_0 x - x_e x) + q(x_e - x) \qquad (13\text{-}8)$$

$$x_e^2 = (p[A]_0 - p x_e - q)x + p[A]_0 x_e + q x_e \qquad (13\text{-}9)$$

42   Reversible Reactions

Equating

$$(p[A]_0 - px_e - q) = 0 \qquad (13\text{-}10)$$

$$p[A]_0 x_e + q x_e = x_e^2 \qquad (13\text{-}11)$$

$$p = \frac{x_e - q}{[A]_0} \qquad (13\text{-}12)$$

Substitution of (13-12) into (13-10) yields q in terms of A_0 and x_e

$$q = \frac{x_e([A]_0 - x_e)}{2[A]_0 - x_e} \qquad (13\text{-}13)$$

This is substituted in turn into equation 13-12 to yield

$$p = \frac{x_e}{2[A]_0 - x_e} \qquad (13\text{-}14)$$

The values for p and q are placed in equation 13-7 and 13-6 to yield

$$\int_0^x \frac{p}{(x_e - x)} dx + \int_0^x \frac{q\, dx}{([A]_0 x_e + [A]_0 x - x_e x)} = k_1 \int_0^t dt \qquad (13\text{-}15)$$

$$\frac{x_e}{(2[A]_0 - x_e)} \int_0^x \frac{dx}{(x_e - x)} + \int_0^x \frac{q\, dx}{[[A]_0 x_e + x([A]_0 - x_e)]} = k_1 \int_0^t dt \qquad (13\text{-}16)$$

$$-\frac{x_e}{(2[A]_0 - x_e)} \int_0^x \frac{d(x_e - x)}{(x_e - x)} + \frac{x_e}{(2[A]_0 - x_e)}$$

$$\times \int_0^x \frac{([A]_0 - x_e)\, dx}{[A]_0 x_e + x([A]_0 - x_e)} = k_1 \int_0^t dt \qquad (13\text{-}17)$$

$$-\frac{x_e}{(2[A]_0 - x_e)} \ln(x_e - x)\Big|_0^x + \frac{x_e}{(2[A]_0 - x_e)}$$

$$\times \int_0^x \frac{d[[A]_0 x_e + x([A]_0 - x_e)]}{[[A]_0 x_e + x([A]_0 - x_e)]} = k_1 t \qquad (13\text{-}18)$$

$$-\frac{x_e}{(2[A]_0 - x_e)} \ln(x_e - x)\Big|_0^x + \frac{x_e}{(2[A]_0 - x_e)}$$

$$\times \ln[[A]_0 x_e + x([A]_0 - x_e)]\Big|_0^x = k_1 t \qquad (13\text{-}19)$$

System 14. Second- and First-order Reversible Reactions

hence

$$k_1 = \frac{x_e}{(2[A]_0 - x_e)t} \ln\left[\frac{[A]_0 x_e + x([A]_0 - x_e)}{[A]_0(x_e - x)}\right] \quad (13\text{-}20)$$

REFERENCES

1. K. Ziegler, L. Ewald, and Ph. Orth, *Annalen*, **479**, 277 (1930).
2. K. Ziegler, Ph. Orth, and K. Weber, *Annalen*, **504**, 131 (1933).
3. E. A. Moelwyn-Hughes, *Trans. Faraday Soc.*, **35**, 368 (1939).

SYSTEM 14. SECOND- AND FIRST-ORDER REVERSIBLE REACTIONS

$$A + B \underset{k_2}{\overset{k_1}{\rightleftharpoons}} C \quad (I)$$

At time $t = 0$ $[A]_0$ $[B]_0$ 0
At time $t = t$ $([A]_0 - x)$ $([B]_0 - x)$ x

A mathematical description of this mechanism is simple if the experimental conditions are such that the initial concentrations of A and B are identical ($[A]_0 = [B]_0$) and that of $[C]$ is zero, then

$$\frac{dx}{dt} = k_1([A]_0 - x)^2 - k_2(x) \quad (14\text{-}1)$$

When the system reaches equilibrium $dx/dt = 0$ and

$$k_1([A]_0 - x_e)^2 = k_2(x_e) \quad (14\text{-}2)$$

$$k_2 = \left[\frac{k_1([A]_0 - x)^2}{(x_e)}\right] \quad (14\text{-}3)$$

where x_e represents the concentration of $[C]$ at equilibrium. Substituting the expression for k_2 into equation 14-1 yields

$$k_1 \, dt = \frac{(x_e) \, dx}{([A]_0 - x)^2(x_e) - ([A]_0 - x_e)^2(x)} \quad (14\text{-}4)$$

$$\frac{(x_e) \, dx}{(x_e - x)([A]_0^2 - x_e x)} = k_1 \, dt \quad (14\text{-}5)$$

44   Reversible Reactions

Equation 14-5 can be integrated after resolution into partial fractions:

$$\frac{x_e}{(x_e - x)([A]_0^2 - x_e x)} = \frac{p}{(x_e - x)} + \frac{q}{([A]_0^2 - x_e x)} \quad (14\text{-}6)$$

$$x_e = p([A]_0^2 - x_e x) + q(x_e - x) \quad (14\text{-}7)$$

$$x_e = p[A]_0^2 + q x_e - x(p x_e + q) \quad (14\text{-}8)$$

$$x_e = p[A]_0^2 + q x_e \quad (14\text{-}9)$$

$$p x_e + q = 0 \quad \text{or} \quad p = \frac{-q}{x_e} \quad (14\text{-}10)$$

$$x_e = p[A]_0^2 - p x_e^2 \quad (14\text{-}11)$$

Hence

$$p = \frac{x_e}{([A]_0^2 - x_e^2)} \quad (14\text{-}12)$$

$$q = -\frac{x_e^2}{([A]_0^2 - x_e^2)} \quad (14\text{-}13)$$

Substitution of the partial fractions and the expressions found for p and q into equation 14-5 yields

$$\frac{x_e}{([A]_0^2 - x_e^2)} \int_0^x \frac{dx}{(x_e - x)} - \frac{x_e}{([A]_0^2 - x_e^2)} \int_0^x \frac{dx}{([A]_0^2 - x_e x)} = k_1 \int_0^t dt \quad (14\text{-}14)$$

$$-\frac{x_e}{([A]_0^2 - x_e^2)} \int_0^x \frac{d(x_e - x)}{(x_e - x)} + \frac{x_e}{([A]_0^2 - x_e^2)} \int_0^x \frac{d([A]_0^2 - x_e x)}{([A]_0^2 - x_e x)} = k_1 \int_0^t dt \quad (14\text{-}15)$$

$$-\frac{x_e}{([A]_0^2 - x_e)} \ln(x_e - x) \Big|_0^x + \frac{x_e}{([A]_0^2 - x_e^2)} \times \ln([A]_0^2 - x_e x) \Big|_0^x = k_1 \int_0^t dt \quad (14\text{-}16)$$

$$-\frac{x_e}{([A]_0^2 - x_e^2)} \ln\left[\frac{(x_e - x)}{x_e}\right] + \frac{x_e}{([A]_0^2 - x_e^2)} \times \ln\left[\frac{([A]_0^2 - x_e x)}{[A]_0^2}\right] = k_1 t \quad (14\text{-}17)$$

System 14. Second- and First-order Reversible Reactions

Hence
$$k_1 = \frac{1}{t} \frac{x_e}{([A]_0^2 - x_e^2)} \ln\left[\frac{x_e([A]_0^2 - x_e x)}{[A]_0^2(x_e - x)}\right] \qquad (14\text{-}18)$$

REFERENCES

1. J. Walker and F. J. Hambly, *Trans. Chem. Soc.* (*London*), **67,** 746 (1895).
2. J. Walker and J. R. Appleyard, *J. Chem. Soc.*, **67,** 193 (1896).
3. G. J. Burrows and E. W. Fawcett, *Trans. Chem. Soc.* (*London*), **105,** 609 (1914).

9

Consecutive Irreversible Reactions

SYSTEM 15. FIRST-ORDER CONSECUTIVE-IRREVERSIBLE THREE-STAGE REACTION

$$A \xrightarrow{k_1} B \xrightarrow{k_2} C \xrightarrow{k_3} D \qquad (I)$$

Amount at time $t = 0$ $\quad [A]_0 \quad\quad 0 \quad\quad 0 \quad\quad 0$
Amount at time $t = t$ $\quad [A] \quad\quad [B] \quad [C] \quad [D]$

This mechanism can be described by the following set of equations:

$$-\frac{d[A]}{dt} = k_1[A] \qquad (15\text{-}1)$$

$$-\frac{d[B]}{dt} = k_1[A] - k_2[B] \qquad (15\text{-}2)$$

$$-\frac{d[C]}{dt} = k_2[B] - k_3[C] \qquad (15\text{-}3)$$

$$\frac{d[D]}{dt} = k_3[C] \qquad (15\text{-}4)$$

These can be solved by two different mathematical operations, the so-called classical method and the operator method.

System 15. Consecutive-irreversible Three-stage Reaction

The Classical Method

Equation 15-1, which represents the rate of disappearance of A, gives after integration

$$[A] = [A]_0 e^{-k_1 t} \tag{15-5}$$

The concentration of B is computed from equation 15-2, which is a linear differential equation of first order. Written in the more familiar form

$$\frac{d[B]}{dt} + k_2[B] = k_1[A] = k_1[A]_0 e^{-k_1 t} \tag{15-6}$$

it is easily identified with the general equation

$$\frac{dy}{dt} + P(x)y = Q(x) \tag{15-7}$$

To solve this equation, the left-hand side has to be transformed into an exact differential. This is done by multiplying both sides of equation 15-6 by the integrating factor

$$\mu(x) = e^{\int P(x)\, dx} \tag{15-8}$$

where $P(x)\, dx$ is equal to $k_2\, dt$, and $\mu(x) = e^{k_2 t}$.

Substitution yields

$$e^{k_2 t} \frac{d[B]}{dt} + e^{k_2 t} k_2 [B] = k_1 [A]_0 e^{-k_1 t} e^{k_2 t} \tag{15-9}$$

$$e^{k_2 t} d[B] + e^{k_2 t} [B] k_2\, dt = k_1 [A]_0 e^{(k_2 - k_1)t}\, dt \tag{15-10}$$

The left-hand side of equation 15-10 is of the form $d(uv) = u\, dv + v\, du$, which is the exact differential $d(e^{k_2 t}[B])$. Equation 15-10 now takes on a form which can be integrated

$$\int_0^{[B]} d(e^{k_2 t}[B]) = k_1 [A]_0 \int_0^t e^{(k_2 - k_1)t}\, dt \tag{15-11}$$

Since the right-hand side of equation 15-11 is of the form

$$\int e^x\, dx = e^x \tag{15-12}$$

48 Consecutive Irreversible Reactions

writing $(k_2 - k_1)t = x$ yields $t = x/(k_2 - k_1)$ and $dt = dx/(k_2 - k_1)$.

Substitution of these expressions into (15-11) gives

$$\int_0^B d(e^{k_2 t}[B]) = \left[\frac{k_1[A]_0}{(k_2 - k_1)}\right]\int_0^t e^{(k_2-k_1)t}\, d(k_2 - k_1)t \quad (15\text{-}13)$$

$$e^{k_2 t}[B] = \left[\frac{k_1[A]_0}{(k_2 - k_1)}\right]\left[e^{(k_2-k_1)t} - 1\right] \quad (15\text{-}14)$$

Hence

$$[B] = \left[\frac{k_1[A]_0}{(k_2 - k_1)}\right]\left[e^{-k_1 t} - e^{-k_2 t}\right] \quad (15\text{-}15)$$

Substitution for $[B]$ into equation 15-3 yields

$$\frac{d[C]}{dt} + k_3[C] = \left[\frac{k_1 k_2[A]_0}{(k_2 - k_1)}\right]\left[e^{-k_1 t} - e^{-k_2 t}\right] \quad (15\text{-}16)$$

The integrating factor in this case is $\mu(x) = e^{\int k_3 dt} = e^{k_3 t}$. Upon multiplying both sides of equation 15-16 by $e^{k_3 t}$, the left-hand side becomes an exact differential $d(e^{k_3 t}[C])$. Hence

$$e^{k_3 t}\,d[C] + k_3 e^{k_3 t}[C]\,dt = \left[\frac{k_1 k_2[A]_0}{(k_2 - k_1)}\right]\left[e^{-k_1 t}e^{k_3 t}\right]$$

$$-\left[\frac{k_1 k_2[A]_0}{(k_2 - k_1)}\right]\left[e^{-k_2 t}e^{k_3 t}\right] \quad (15\text{-}17)$$

$$\int_0^t d(e^{k_3 t}[C]) = \left[\frac{k_1 k_2[A]_0}{(k_2 - k_1)}\right]\int_0^t [e^{(k_3-k_1)t} - e^{(k_3-k_2)t}]\, dt \quad (15\text{-}18)$$

An analysis of the left-hand side of equation 15-18 shows that, since at $t = 0$, $C = 0$,

$$\int_0^t d(e^{k_3 t}[C]) = e^{k_3 t}[C] \quad (15\text{-}19)$$

System 15. Consecutive-irreversible Three-stage Reaction

The right-hand side of equation 15-18 can be written

$$\left[\frac{k_1 k_2 [A]_0}{(k_2 - k_1)}\right]\int_0^t e^{(k_3-k_1)t} \, dt - \left[\frac{k_1 k_2 [A]_0}{(k_2 - k_1)}\right]\int_0^t e^{(k_3-k_2)t} \, dt \quad (15\text{-}20)$$

and is treated in the same manner as equation 15-11. Hence when integrated, equation 15-18 becomes

$$[C]e^{k_3 t} = \left[\frac{k_1 k_2 [A]_0}{(k_2 - k_1)(k_3 - k_1)}\right]\left[e^{(k_3-k_1)t} - 1\right]$$
$$- \left[\frac{k_1 k_2 [A]_0}{(k_2 - k_1)(k_3 - k_2)}\right]\left[e^{(k_3-k_2)t} - 1\right] \quad (15\text{-}19)$$

Hence

$$[C] = \left[\frac{k_1 k_2 [A]_0}{(k_2 - k_1)(k_3 - k_1)}\right]e^{-k_1 t} + \left[\frac{k_1 k_2 [A]_0}{(k_1 - k_2)(k_3 - k_2)}\right]e^{-k_2 t}$$
$$+ \left[\frac{k_1 k_2 [A]_0}{(k_1 - k_3)(k_2 - k_3)}\right]e^{-k_3 t} \quad (15\text{-}20)$$

The concentration of product D, can be determined from the material balance equation

$$[D] = [A]_0 - [A] - [B] - [C] \quad (15\text{-}21)$$

substitution of the values obtained from equations 15-5, 15-15, and 15-20 gives

$$[D] = [A]_0\left[1 - \frac{k_2 k_3 e^{-k_1 t}}{(k_2 - k_1)(k_3 - k_1)} - \frac{k_1 k_3 e^{-k_2 t}}{(k_1 - k_2)(k_3 - k_2)}\right.$$
$$\left. - \frac{k_1 k_2 e^{-k_3 t}}{(k_1 - k_3)(k_2 - k_3)}\right] \quad (15\text{-}22)$$

The Operator Method

The linear differential equations 15-1, 15-2, 15-3, and 15-4 are of such form that they can be solved by the operator method.

50 Consecutive Irreversible Reactions

In applying this procedure one replaces differentiation by multiplication, in which the multiplier is operator $S = d/dt$. In addition, unknown functions in the system of differential equations are replaced by their corresponding transforms.

Substitution of $[A]_0$ for $[A]$ at time zero, when $[B]$, $[C]$, and $[D]$ are equal to zero, yields

$$\frac{d([A]_0 - [A])}{dt} = S[A]_0 - S[A] = k_1[A] \quad (15\text{-}23)$$

hence

$$[A] = \frac{S[A]_0}{(S + k_1)} \quad (15\text{-}24)$$

From tables of Laplace transforms one finds that

$$\frac{S}{(S \pm a)} = e^{\pm at} \quad \text{so that} \quad [A] = [A]_0 e^{-k_1 t} \quad (15\text{-}25)$$

The other differential equations can be solved in a similar manner: For

$$\frac{d[B]}{dt} = S[B] = k_1[A] - k_2[B] \quad (15\text{-}26)$$

$$[B] = \frac{k_1[A]_0 S}{(S + k_1)(S + k_2)} \quad (15\text{-}27)$$

$$[B] = [A]_0 \left[\frac{k_1}{(k_2 - k_1)} e^{-k_1 t} + \frac{k_1}{(k_1 - k_2)} e^{-k_2 t} \right] \quad (15\text{-}28)$$

For

$$\frac{d[C]}{dt} = S[C] = k_2[B] - k_3[C] \quad (15\text{-}29)$$

$$[C] = \frac{k_2[B]}{(S + k_3)} \quad (15\text{-}30)$$

Substitution for $[B]$ gives

$$[C] = \frac{k_1 k_2 [A]_0 S}{(S + k_1)(S + k_2)(S + k_3)} \quad (15\text{-}31)$$

System 15. Consecutive-irreversible Three-stage Reaction

Hence the final form for C is

$$[C] = [A]_0 \left[\frac{k_1 k_2}{(k_2 - k_1)(k_3 - k_1)} e^{-k_1 t} + \frac{k_1 k_2}{(k_1 - k_2)(k_3 - k_2)} e^{-k_2 t} \right.$$
$$\left. + \frac{k_1 k_2}{(k_1 - k_3)(k_2 - k_3)} e^{-k_3 t} \right] \quad (15\text{-}34)$$

Similarly,

$$\frac{d[D]}{dt} = S[D] = k_3[C]$$

hence
$$[D] = \frac{k_1 k_2 k_3 [A]_0}{(S + k_1)(S + k_2)(S + k_3)} \quad (15\text{-}35)$$

$$[D] = [A]_0 \left[1 - \frac{k_2 k_3}{(k_2 - k_1)(k_3 - k_1)} e^{-k_1 t} - \frac{k_1 k_3}{(k_1 - k_2)(k_3 - k_2)} \right.$$
$$\left. \times e^{-k_2 t} - \frac{k_1 k_2}{(k_1 - k_3)(k_2 - k_3)} e^{-k_3 t} \right] \quad (15\text{-}36)$$

REFERENCES

1. N. Esson, *Phil. Trans. Roy. Soc. (London)*, **220**, 156 (1866).
2. A. Rakowsky, *Z. Phys. Chem.*, **57**, 321 (1906).
3. F. Tiersch, *Z. Phys. Chem.*, **111**, 175 (1924).
4. H. Clement and I. Sovard, *Compt. Rend.*, **206**, 610 (1938).
5. H. Clement, *Ann. Chimie*, **13**, 243 (1940).
6. B. V. Erofeev, *Z. Fiz. Khim.*, **24**, 721 (1950).
7. A. Westman and D. B. DeLury, *Can. J. Chem.*, **34**, 1134 (1956).
8. V. G. Plyusnin and N. M. Rodiguin, *Zh. Fiz. Khim.*, **31**, 2066 (1957).
9. V. G. Plyusnin, A. P. Lysenko, and Ye. P. Babin, *Zh. Fiz. Khim.*, **31**, 2229 (1957).
10. A. P. Lysenko and V. G. Plyusnin, *Zh. Fiz. Khim.*, **32**, 1074 (1958).

SYSTEM 16. FIRST-ORDER CONSECUTIVE-IRREVERSIBLE $n-1$ STAGE REACTION

$$A_1 \xrightarrow{k_1} A_2 \xrightarrow{k_2} A_3 \cdots A_{n-2} \xrightarrow{k_{n-2}} A_{n-1} \xrightarrow{k_{n-1}} A_n \quad (I)$$

Amount at time $t = 0$: $[A_1]_0$, 0, 0, 0, 0, 0

Amount at time $t = t$: $[A_1]$, $[A_2]$, $[A_3]$, $[A_{n-2}]$, $[A_{n-1}]$, $[A_n]$

Starting with a single substance $[A_1]_0$ at time $t = 0$, the reaction runs through $(n-1)$ stages. The objective is to derive expressions that will describe the concentrations of the intermediates existing at any time t during the reaction. The set of differential equations (16-1 to 5), which describes this mechanism, is solved here by two different mathematical methods.

$$\frac{d[A_1]}{dt} = -k_1[A_1] \tag{16-1}$$

$$\frac{d[A_2]}{dt} = k_1[A_1] - k_2[A_2] \tag{16-2}$$

$$\frac{d[A_3]}{dt} = k_2[A_2] - k_3[A_3] \tag{16-3}$$

$$\cdots$$
$$\cdots$$

$$\frac{d[A_{n-1}]}{dt} = k_{n-2}[A_{n-2}] - k_{n-1}[A_{n-1}] \tag{16-4}$$

$$\frac{d[A_n]}{dt} = k_{n-1}[A_{n-1}] \tag{16-5}$$

System 16. Consecutive-irreversible $n - 1$ Stage Reaction

The Classical Method

The systematic derivation of a more general equation for A_{n-1} stages is illustrated on the mathematical derivations for products A_3 and A_4. As will be recalled from System 15, the concentration of the second product is described by equation 15-34

$$[A_3] = \{p_3\}_1 e^{-k_1 t} + \{p_3\}_2 e^{-k_2 t} + \{p_3\}_3 e^{-k_3 t} \tag{16-6}$$

where

$$\{p_3\}_1 = \left[\frac{k_1 k_2 [A_1]_0}{(k_2 - k_1)(k_3 - k_1)}\right] \tag{16-7}$$

$$\{p_3\}_2 = \left[\frac{k_1 k_2 [A_1]_0}{(k_1 - k_2)(k_3 - k_2)}\right] \tag{16-8}$$

$$\{p_3\}_3 = \left[\frac{k_1 k_2 [A_1]_0}{(k_3 - k_1)(k_3 - k_2)}\right] \tag{16-9}$$

The rate expression for $[A_4]$ is given by

$$\frac{d[A_4]}{dt} = k_3 [A_3] - k_4 [A_4] \tag{16-10}$$

or replacing A_3 by equation 16-6

$$d[A_4] + k_4 [A_4]\, dt = k_3 [\{p_3\}_1 e^{-k_1 t} + \{p_3\}_2 e^{-k_2 t} + \{p_3\}_3 e^{-k_3 t}]\, dt \tag{16-11}$$

Equation 16-11 is multiplied by the itegrating factor $\mu = e^{\int k_4 dt} = e^{k_4 t}$, and integrated

$$\int_0^t d(e^{k_4 t} [A_4]) = \{p_3\}_1 k_3 \int_0^t e^{(k_4 - k_1)t}\, dt + \{p_3\}_2 k_3 \int_0^t e^{(k_4 - k_2)t}\, dt$$

$$+ \{p_3\}_3 k_3 \int_0^t e^{(k_4 - k_3)t} dt \tag{16-12}$$

54 Consecutive Irreversible Reactions

$$e^{k_4 t}[A_4] = \left(\frac{\{p_3\}_1 k_3}{(k_4 - k_1)}\right)[e^{(k_4-k_1)t} - 1] + \left(\frac{\{p_3\}_2 k_3}{(k_4 - k_2)}\right)$$

$$\times \left[e^{(k_4-k_2)t} - 1\right] + \left(\frac{\{p_3\}_3 k_3}{(k_4 - k_3)}\right)[e^{(k_4-k_3)t} - 1] \quad (16\text{-}13)$$

$$[A_4] = \left(\frac{k_1 k_2 k_3 [A_1]_0 e^{-k_1 t}}{(k_2 - k_1)(k_3 - k_1)(k_4 - k_1)}\right)$$

$$+ \left(\frac{k_1 k_2 k_3 [A_1]_0 e^{-k_2 t}}{(k_1 - k_2)(k_3 - k_2)(k_4 - k_2)}\right)$$

$$+ \left(\frac{k_1 k_2 k_3 [A_1]_0 e^{-k_3 t}}{(k_1 - k_3)(k_2 - k_3)(k_4 - k_3)}\right)$$

$$- k_1 k_2 k_3 [A_1]_0 e^{-k_4 t} \left[\frac{1}{(k_2 - k_1)(k_3 - k_1)(k_4 - k_1)}\right.$$

$$+ \frac{1}{(k_1 - k_2)(k_3 - k_2)(k_4 - k_2)}$$

$$\left. + \frac{1}{(k_1 - k_3)(k_2 - k_3)(k_4 - k_3)}\right] \quad (16\text{-}14)$$

$$[A_4] = \{p_4\}_1 e^{-k_1 t} + \{p_4\}_2 e^{-k_2 t} + \{p_4\}_3 e^{-k_3 t}$$

$$+ \frac{k_1 k_2 k_3 [A_1]_0 e^{-k_4 t}}{(k_1 - k_4)(k_2 - k_4)(k_3 - k_4)} \left[\frac{(k_2 - k_4)(k_3 - k_4)}{(k_2 - k_1)(k_3 - k_1)}\right.$$

$$\left. + \frac{(k_1 - k_4)(k_3 - k_4)}{(k_1 - k_2)(k_3 - k_2)} + \frac{(k_1 - k_4)(k_2 - k_4)}{(k_1 - k_3)(k_2 - k_3)}\right] \quad (16\text{-}15)$$

or

$$[A_4] = \{p_4\}_1 e^{-k_1 t} + \{p_4\}_2 e^{-k_2 t} + \{p_4\}_3 e^{-k_3 t} + \{p_4\}_4 e^{-k_4 t} \quad (16\text{-}16)$$

System 16. Consecutive-irreversible $n - 1$ Stage Reaction

where

$$\{p_4\}_1 = \left[\frac{k_1 k_2 k_3 [A_1]_0}{(k_2 - k_1)(k_3 - k_1)(k_4 - k_1)}\right] \quad (16\text{-}17)$$

$$\{p_4\}_2 = \left[\frac{k_1 k_2 k_3 [A_1]_0}{(k_1 - k_2)(k_3 - k_2)(k_4 - k_2)}\right] \quad (16\text{-}18)$$

$$\{p_4\}_3 = \left[\frac{k_1 k_2 k_3 [A_1]_0}{(k_1 - k_3)(k_2 - k_3)(k_4 - k_3)}\right] \quad (16\text{-}19)$$

$$\{p_4\}_4 = \left[\frac{k_1 k_2 k_3 [A_1]_0}{(k_1 - k_4)(k_2 - k_4)(k_3 - k_4)}\right] \quad (16\text{-}20)$$

The form of equations 16-6 to 16-9 and 16-16 to 16-20 suggests that a more general equation can be written for the concentration of product A_{n-1},

$$[A_{n-1}] = \{p_{n-1}\}_1 e^{-k_1 t} + \{p_{n-1}\}_2 e^{-k_2 t} + \cdots + \{p_{n-1}\}_{n-1} e^{-k_{n-1} t}$$

$$(16\text{-}21)$$

where the terms $\{p_{n-1}\}_1, \{p_{n-1}\}_2, \ldots, \{p_{n-1}\}_{n-1}$ are

$$\{p_{n-1}\}_1 = \left[\frac{k_1 k_2 k_3 k_4 \cdots k_{n-2} [A_1]_0}{(k_2 - k_1)(k_3 - k_1) \cdots (k_{n-1} - k_1)}\right] \quad (16\text{-}22)$$

$$\{p_{n-1}\}_2 = \left[\frac{k_1 k_2 k_3 k_4 \cdots k_{n-2} [A_1]_0}{(k_1 - k_2)(k_3 - k_2) \cdots (k_{n-1} - k_2)}\right] \quad (16\text{-}23)$$

$$\{p_{n-1}\}_3 = \left[\frac{k_1 k_2 k_3 k_4 \cdots k_{n-2} [A_1]_0}{(k_1 - k_3)(k_2 - k_3) \cdots (k_{n-1} - k_3)}\right] \quad (16\text{-}24)$$

$$\cdots$$
$$\cdots$$

$$\{p_{n-1}\}_{n-1} = \left[\frac{k_1 k_2 k_3 k_4 \cdots k_{n-2} [A_1]_0}{(k_1 - k_{n-1})(k_2 - k_{n-1}) \cdots (k_{n-2} - k_{n-1})}\right]$$

$$(16\text{-}25)$$

56 Consecutive Irreversible Reactions

To derive the general equation for the final product A_n, one can assume for convenience that the reaction scheme involves one more stage:

$$A_1 \xrightarrow{k_1} A_2 \xrightarrow{k_2} A_3 \cdots \xrightarrow{k_{n-1}} A_n \xrightarrow{k_n} A_{n+1} \qquad \text{(II)}$$

$$\begin{aligned}[A_n] &= \left[\frac{k_1 k_2 k_3 k_4 \cdots k_{n-1}[A_1]_0 e^{-k_1 t}}{(k_2 - k_1)(k_3 - k_1) \cdots (k_n - k_1)}\right] \\ &+ \left[\frac{k_1 k_2 k_3 k_4 \cdots k_{n-1}[A_1]_0 e^{-k_2 t}}{(k_1 - k_2)(k_3 - k_2) \cdots (k_n - k_2)}\right] \\ &+ \left[\frac{k_1 k_2 k_3 k_4 \cdots k_{n-1}[A_1]_0 e^{-k_3 t}}{(k_1 - k_3)(k_2 - k_3) \cdots (k_n - k_3)}\right] \\ & \cdots \\ &+ \left[\frac{k_1 k_2 k_3 k_4 \cdots k_{n-2} k_{n-1}[A_1]_0 e^{-k_{n-1} t}}{(k_1 - k_{n-1})(k_2 - k_{n-1}) \cdots (k_{n-2} - k_{n-1})(k_n - k_{n-1})}\right] \\ &+ \left[\frac{k_1 k_2 k_3 k_4 \cdots k_{n-1}[A_1]_0 e^{-k_n t}}{(k_1 - k_n)(k_2 - k_n) \cdots (k_{n-1} - k_n)}\right] \qquad (16\text{-}26)\end{aligned}$$

If A_n is indeed the final product, then $k_n = 0$, hence

$$\begin{aligned}[A_n] = [A_1]_0 &\bigg[1 - \left(\frac{k_2 k_3 k_4 \cdots k_{n-1} e^{-k_1 t}}{(k_2 - k_1)(k_3 - k_1) \cdots (k_{n-1} - k_1)}\right) \\ &- \left(\frac{k_1 k_3 k_4 \cdots k_{n-1} e^{-k_2 t}}{(k_1 - k_2)(k_3 - k_2) \cdots (k_{n-1} - k_2)}\right) \\ &- \left(\frac{k_1 k_2 k_4 \cdots k_{n-1} e^{-k_3 t}}{(k_1 - k_3)(k_2 - k_3) \cdots (k_{n-1} - k_3)}\right) \\ &\cdots \\ &- \left(\frac{k_1 k_2 k_3 k_4 \cdots k_{n-2} e^{-k_{n-1} t}}{(k_1 - k_{n-1})(k_2 - k_{n-1}) \cdots (k_{n-2} - k_{n-1})}\right)\bigg] \\ &= [A_1]_0 - \{[A_1] + [A_2] + \cdots + [A_{n-1}]\} \qquad (16\text{-}27)\end{aligned}$$

System 16. Consecutive-irreversible $n - 1$ Stage Reaction　　57

The Operator Method

The unknown functions in equations 16-1 to 16-5 are replaced by their corresponding transforms (See Appendix II)

$$S[A_1] - S[A_1]_0 = -k_1[A_1] \tag{16-28}$$

$$S[A_2] = k_1[A_1] - k_2[A_2] \tag{16-29}$$

$$S[A_3] = k_2[A_2] - k_3[A_3] \tag{16-30}$$

$$\cdots$$

$$S[A_{n-1}] = k_{n-2}[A_{n-2}] - k_{n-1}[A_{n-1}] \tag{16-31}$$

$$S[A_n] = k_{n-1}[A_{n-1}] \tag{16-32}$$

The solutions to these equations are

$$[A_1] = \frac{S[A_1]_0}{(S + k_1)} \tag{16-33}$$

$$[A_2] = \frac{k_1 S[A_1]_0}{(S + k_1)(S + k_2)} \tag{16-34}$$

$$[A_3] = \frac{k_1 k_2 S[A_1]_0}{(S + k_1)(S + k_2)(S + k_3)} \tag{16-35}$$

$$\cdots$$

$$[A_{n-1}] = \frac{k_1 k_2 \cdots k_{n-2} S[A_1]_0}{(S + k_1)(S + k_2) \cdots (S + k_n)} \tag{16-36}$$

$$[A_n] = \frac{k_1 k_2 \cdots k_{n-1} S[A_1]_0}{(S + k_1)(S + k_2) \cdots (S + k_n)} \tag{16-37}$$

Next the transforms of the unknown functions $A_1, A_2 \cdots A_n$ are replaced by their originals which are tabulated in Appendix II:

$$[A_1] = [A_1]_0 e^{-k_1 t} \tag{16-38}$$

$$[A_2] = [A_1]_0 \left(\frac{k_1}{k_2 - k_1} e^{-k_1 t} + \frac{k_1}{k_1 - k_2} e^{-k_2 t} \right) \tag{16-39}$$

58 Consecutive Irreversible Reactions

$$[A_3] = [A_1]_0 \left\{ \left[\frac{k_1 k_2 e^{-k_1 t}}{(k_2 - k_1)(k_3 - k_1)} \right] + \left[\frac{k_1 k_2 e^{-k_2 t}}{(k_1 - k_2)(k_3 - k_2)} \right] \right.$$
$$\left. + \left[\frac{k_1 k_2 e^{-k_3 t}}{(k_1 - k_3)(k_2 - k_3)} \right] \right\} \quad (16\text{-}40)$$

$$\cdots$$
$$\cdots$$

$$[A_{n-1}] = [A_1]_0 \left\{ \left[\frac{k_1 k_2 \cdots k_{n-2} e^{-k_1 t}}{(k_2 - k_1)(k_3 - k_1) \cdots (k_{n-1} - k_1)} \right] \right.$$
$$+ \left[\frac{k_1 k_2 \cdots k_{n-2} e^{-k_2 t}}{(k_1 - k_2)(k_3 - k_2) \cdots (k_{n-1} - k_2)} \right] + \cdots$$
$$+ \left[\frac{k_1 k_2 \cdots k_{n-2} e^{-k_{n-1} t}}{(k_1 - k_{n-1})(k_2 - k_{n-1})(k_3 - k_{n-1}) \cdots (k_{n-2} - k_{n-1})} \right]$$
$$(16\text{-}41)$$

$$[A_n] = [A_1]_0 \left\{ 1 - \left[\frac{k_2 k_3 \cdots k_{n-1} e^{-k_1 t}}{(k_2 - k_1)(k_3 - k_1) \cdots (k_{n-1} - k_1)} \right] \right.$$
$$+ \left[\frac{k_1 k_3 \cdots k_{n-1} e^{-k_2 t}}{(k_1 - k_2)(k_3 - k_2) \cdots (k_{n-1} - k_2)} \right]$$
$$\cdots$$
$$- \left[\frac{k_1 k_2 k_4 \cdots k_{n-2} e^{-k_3 t}}{(k_1 - k_3)(k_2 - k_3) \cdots (k_{n-2} - k_3)} \right]$$
$$\cdots$$
$$- \left[\frac{k_1 k_2 k_3 \cdots k_{n-2} e^{-k_{n-1} t}}{(k_1 - k_{n-1})(k_2 - k_{n-1}) \cdots (k_{n-2} - k_{n-1})} \right]$$
$$= [A_1]_0 - \{[A_1] + [A_2] + [A_3] + \cdots + [A_{n-1}]\} \quad (16\text{-}42)$$

REFERENCES

1. E. Abel, *Z. Phys. Chem.*, **A56,** 558 (1906).
2. H. Batemen, *Proc. Camb. Phil. Soc.*, **15,** 423 (1910).
3. B. V. Yerofeyev, *Zh. Fiz. Khim.*, **24,** 721 (1950).
4. A. N. Murin, *Introduction to Radioactivity*, Izd. Leningrad State University, 1955.
5. V. G. Plyusnin and N. M. Rodiguin, *Zh. Fiz. Khim.*, **31,** 2066 (1957).

10

Catalytic Reactions

SYSTEM 17. SECOND-ORDER AUTOCATALYTIC REACTION. TYPE I.

When a product of a chemical reaction catalyzes (speeds up) decomposition of the reactant(s), the process is said to be autocatalytic. Typical examples of such kinetics are the acid-catalyzed hydrolyses of esters. For example, the hydrolysis of methyl acetate can be initiated by the addition of a small amount of acetic acid; as the reaction progresses and more acetic acid is produced, the rate of ester hydrolysis will increase proportionately.

In the reaction

$$\text{RCOOH} + \text{RCOOR}' + \text{H}_2\text{O} \xrightarrow{k_1} 2\text{RCOOH} + \text{R}'\text{OH} \qquad \text{(I)}$$

the initial concentrations of the ester (RCOOR′) and the acid (RCOOH) are represented by $[A]_0$ and $[B]_0$, respectively, and if x is the amount of ester hydrolyzed in time t, the rate of reaction at any instant is given by

$$\frac{dx}{dt} = k_1(\text{H}_2\text{O})([A]_0 - x)([B]_0 + x) \qquad (17\text{-}1)$$

$$\frac{dx}{dt} = k'([A]_0 - x)([B]_0 + x) \qquad (17\text{-}2)$$

60 Catalytic Reactions

The concentration of water is effectively constant and can be included in the rate constant k'. It is assumed that the rate is proportional to concentration of the acetic acid catalyst. Equation 17-2 can be integrated after resolution into partial fractions:

$$\int_0^x \frac{dx}{([A]_0 - x)([B]_0 + x)} = k' \int_0^t dt \qquad (17\text{-}3)$$

Resolution into partial fractions:

$$\frac{1}{([A]_0 - x)([B]_0 + x)} = \frac{p}{([A]_0 - x)} + \frac{q}{([B]_0 + x)} \qquad (17\text{-}4)$$

$$1 = p([B]_0 + x) + q([A]_0 - x) \qquad (17\text{-}5)$$

$$1 = p[B]_0 + px + q[A]_0 - qx \qquad (17\text{-}6)$$

$$1 = p[B]_0 + q[A]_0 + x(p - q) \qquad (17\text{-}7)$$

Setting

$$p[B]_0 + q[A]_0 = 1 \qquad (17\text{-}8)$$

and

$$p - q = 0, \quad p = q \qquad (17\text{-}9)$$

hence

$$p[A]_0 + p[B]_0 = 1 \qquad (17\text{-}10)$$

or

$$p = \frac{1}{([A]_0 + [B]_0)} = q \qquad (17\text{-}11)$$

Substitution of the resolved fractions and the expressions derived for the constants p and q into equation 17-3 yields a differential equation which can be integrated as follows:

$$\int_0^x \left[\frac{p}{([A]_0 - x)} + \frac{q}{([B]_0 + x)} \right] dx = k' \int_0^t dt \qquad (17\text{-}12)$$

$$\frac{1}{([A]_0 + [B]_0)} \int_0^x \frac{dx}{([A]_0 - x)} + \frac{1}{([B]_0 + [A]_0)} \int_0^x \frac{dx}{([B]_0 + x)} = k' \int_0^t dt$$

$$(17\text{-}13)$$

System 17. Second-order Autocatalytic Reaction. Type I

$$-\frac{1}{([B]_0 + [A]_0)} \int_0^x \frac{d([A]_0 - x)}{([A]_0 - x)}$$

$$+ \frac{1}{([B]_0 + [A]_0)} \int_0^x \frac{d([B]_0 + x)}{([B]_0 + x)} = k' \int_0^t dt \quad (17\text{-}14)$$

$$\frac{1}{([B]_0 + [A]_0)} [\ln [A]_0 - \ln ([A]_0 - x)]$$

$$+ \frac{1}{([B]_0 + [A]_0)} [\ln ([B]_0 + x) - \ln [B]_0] = k't \quad (17\text{-}15)$$

hence

$$k' = \frac{1}{t([B]_0 + [A]_0)} \ln \left[\frac{[A]_0([B]_0 + x)}{[B]_0([A]_0 - x)}\right] \quad (17\text{-}16)$$

Rearrangement of equation 17-16 leads to an expression which illustrates the variation of $([B]_0 + x)$ with time:

$$([B]_0 + x) = ([A]_0 - x)\left(\frac{[B]_0}{[A]_0}\right) e^{[([A]_0 + [B]_0)k't]} \quad (17\text{-}17)$$

Replacing $([A]_0 - x)$ by its equivalent form $([A]_0 - x) = ([A]_0 + [B]_0) - ([B]_0 + x)$ yields

$$[B]_0 = [([A]_0 + [B]_0) - ([B]_0 + x)]\left(\frac{[B]_0}{[A]_0}\right) e^{[([A]_0 + [B]_0)k't]} \quad (17\text{-}18)$$

or

$$([B]_0 + x) = [([A]_0 + [B]_0) - ([B]_0 + x)]\left(\frac{[B]_0}{[A]_0}\right) e^{[([A]_0 + [B]_0)k't]}$$
$$(17\text{-}19)$$

further rearrangement yields

$$([B]_0 + x)\left[1 + \left(\frac{[B]_0}{[A]_0}\right) e^{[([A]_0 + [B]_0)k't]}\right]$$

$$= ([A]_0 + [B]_0)\left(\frac{[B]_0}{[A]_0}\right) e^{[([A]_0 + [B]_0)k't]} \quad (17\text{-}20)$$

In dividing equation 17-20 by $[[B]_0/[A]_0]e^{([A]_0+[B]_0)k't}$ one obtains the final expression

$$([B]_0 + x) = \frac{[A]_0 + [B]_0}{\{1 + [([A]_0/[B]_0)]e^{-[([A]_0+[B]_0)k't]}\}} \quad (17\text{-}21)$$

SYSTEM 18. SECOND-ORDER AUTOCATALYTIC REACTION. TYPE II

When the catalyst (H^+ as HCl) added initially to the system is different from the one (H^+ as RCOOH) produced in the chemical reaction, as, for example, in acid catalyzed hydrolysis of esters

$$RCOOR' + H_2O \xrightarrow{HCl} RCOO^- + H^+ + R'OH \quad (I)$$

the rate of reaction obeys the following differential equation:

$$\frac{dx}{dt} = k_1([A]_0 - x)[B]_0 + k_2([A]_0 - x)x \quad (18\text{-}1)$$

where $[A]_0$ is the initial concentration of the ester (RCOOR'), $[B]_0$ is the amount of catalyst (HCl) present at the start of the reaction, when $t = 0$, and x is the amount of catalyst $[H^+]$ formed in time $t = t$.

Equation 18-1 can be integrated after rearrangement and resolution into partial fractions:

$$\int_0^x \frac{dx}{k_1([A]_0 - x)[B]_0 + k_2([A]_0 - x)x} = \int_0^t dt \quad (18\text{-}2)$$

$$\int_0^x \frac{dx}{([A]_0 - x)(k_1[B]_0 + k_2x)} = \int_0^t dt \quad (18\text{-}3)$$

Resolution into partial fractions:

$$\frac{1}{([A]_0 - x)(k_1[B]_0 + k_2x)} = \frac{p}{([A]_0 - x)} + \frac{q}{(k_1[B]_0 + k_2x)} \quad (18\text{-}4)$$

$$1 = p(k_1[B]_0 + k_2x) + q([A]_0 - x) \quad (18\text{-}5)$$

$$1 = pk_1[B]_0 + q[A]_0 \quad (18\text{-}6)$$

Setting

$$pk_2 - q = 0 \quad (18\text{-}7)$$
$$pk_2 = q \quad (18\text{-}8)$$

and

$$pk_1[B]_0 + pk_2[A]_0 = 1 \quad (18\text{-}9)$$

yields

$$p = \frac{1}{k_1[B]_0 + k_2[A]_0} \quad (18\text{-}10)$$

$$q = \frac{k_2}{k_1[B]_0 + k_2[A]_0} \quad (18\text{-}11)$$

Substitution of these terms back into equation 18-3 yields

$$\int_0^x \frac{p\,dx}{([A]_0 - x)} + \int_0^x \frac{q\,dx}{(k_1[B]_0 + k_2x)} = \int_0^t dt \quad (18\text{-}12)$$

$$-\frac{1}{(k_1[B]_0 + k_2[A]_0)} \int_0^x \frac{d([A]_0 - x)}{([A]_0 - x)}$$
$$+ \frac{1}{(k_1[B]_0 + k_2[A]_0)} \int_0^x \frac{k_2\,dx}{(k_1[B]_0 + k_2x)} = \int_0^t dt \quad (18\text{-}13)$$

$$\frac{1}{(k_1[B]_0 + k_2[A]_0)} \ln\left[\frac{[A]_0}{([A]_0 - x)}\right]$$
$$+ \frac{1}{(k_1[B]_0 + k_2[A]_0)} \int_0^x \frac{d(k_1[B]_0 + k_2x)}{(k_1[B]_0 + k_2x)} = t \quad (18\text{-}14)$$

$$\frac{1}{(k_1[B]_0 + k_2[A]_0)} \ln\left[\frac{[A]_0}{([A]_0 - x)}\right]$$
$$+ \frac{1}{(k_1[B]_0 + k_2[A]_0)} \ln\left[\frac{k_1[B]_0 + k_2x}{k_1[B]_0}\right] = t \quad (18\text{-}15)$$

$$t = \frac{1}{(k_1[B]_0 + k_2[A]_0)} \left[\ln\left(\frac{[A]_0}{[A]_0 - x}\right) + \ln\left(\frac{(k_1[B]_0 + k_2x)}{k_1[B]_0}\right)\right] \quad (18\text{-}16)$$

or

$$k_1[B]_0 + k_2[A]_0 = \frac{1}{t} \ln\left[\frac{[A]_0(k_1[B]_0 + k_2x)}{k_1[B]_0([A]_0 - x)}\right] \quad (18\text{-}17)$$

SYSTEM 19. AN ACID-BASE CATALYZED DECAY REACTION OF FIRST ORDER

$$H_2O_3 + H^+ \xrightarrow{k_1} H_3O^+ + O_2 \quad \text{(I)}$$
$$H_2O_3 \xrightarrow{k_2} HO_3^- + H^+ \quad \text{(II)}$$
$$HO_3^- + H^+ \xrightarrow{k_3} H_2O_3 \quad \text{(III)}$$
$$HO_3^- \xrightarrow{k_4} OH^- + O_2 \quad \text{(IV)}$$
$$H_2O_3 + OH^- \xrightarrow{k_5} H_2O + HO_3^- \quad \text{(V)}$$

This mechanism describes a situation where the observed rate constant for unimolecular decay of H_2O_3 varies with pH. The rate at which H_3O_3 and its dissociated form HO_3^- disappear is given by

$$-\frac{d[H_2O_3]}{dt} = k_1[H_2O_3][H^+] + k_2[H_2O_3]$$
$$- k_3[HO_3^-][H^+] + k_5[H_2O_3][OH^-] \quad (19\text{-}1)$$

$$-\frac{d[HO_3^-]}{dt} = -k_2[H_2O_3] + k_3[HO_3^-][H^+]$$
$$+ k_4[HO_3^-] - k_5[H_2O_3][OH^-] \quad (19\text{-}2)$$

Assuming steady-state conditions equation 19-2 yields

$$k_2[H_2O_3] + k_5[H_2O_3][OH^-] = k_3[HO_3^-][H^+] + k_4[HO_3^-] \quad (19\text{-}3)$$

hence

$$[HO_3^-] = \frac{[H_2O_3](k_2 + k_5[OH^-])}{(k_4 + k_3[H^+])} \quad (19\text{-}4)$$

Appropriate substitution of terms from equations 19-3 and 19-4 into 19-1 gives the rate equation 19-5, which accounts for the acid-base catalyzed disappearance of H_2O_3:

$$-\frac{d[H_2O_3]}{dt} = \left[k_1[H^+] + \frac{k_2 + k_5[OH^-]}{1 + (k_3/k_4)[H^+]}\right][H_2O_3] \equiv k_{H_2O_3}$$
$$(19\text{-}5)$$

System 20. Catalyzed Decay of a Dissociable Free Radical

A graph of $1/k_{H_2O_3}$ against $[H^+]$ gives in the proper acid range a straight line obeying the equation

$$\frac{1}{k_{H_2O_3}} = a + b[H^+] \tag{19-6}$$

from which the value for k_2 and k_3/k_4 are evaluated. The pK for H_2O_3 is given by $-\log k_2/k_3$

REFERENCES

1. G. Czapski and B. H. J. Bielski, *J. Phys. Chem.*, **67**, 2180 (1963).
2. B. H. J. Bielski and H. A. Schwarz, *J. Phys. Chem.*, **72**, 3836 (1968).
3. B. H. J. Bielski, *J. Phys. Chem.*, **74**, 3213 (1970).

SYSTEM 20. SECOND-ORDER ACID CATALYZED DECAY OF A DISSOCIABLE FREE RADICAL

$$QH\cdot + QH\cdot \xrightarrow{k_1} QH_2 + Q \tag{I}$$

$$QH\cdot \rightleftharpoons Q^- + H^+ \tag{1}$$

$$QH\cdot + Q^- + H^+ \xrightarrow{k_2} QH_2 + Q \tag{II}$$

$$Q^- + Q^- + 2H^+ \xrightarrow{k_3} QH_2 + Q \tag{III}$$

This mechanism describes an acid-dependent disproportionation reaction of second-order, for which the overall rate of disappearance of the radicals is a complex function of its dissociation constant, K,

$$K = \frac{[Q^-][H^+]}{[QH\cdot]} \tag{20-1}$$

Since the total radical concentration R present at any time is

$$[R] = [QH\cdot] + [Q^-] \tag{20-2}$$

Catalytic Reactions

the mechanism above can be described by the following equation:

$$-\frac{d[R]}{dt} = 2k_1[QH\cdot]^2 + 2k_2[QH\cdot][Q^-] + 2k_3[Q^-]^2 \quad (20\text{-}3)$$

This equation can be solved upon substitution of expressions (20-4) and (20-5) derived from the dissociation constant and equation (20-2),

$$[QH\cdot] = \frac{R}{(1 + K/H^+)} \quad (20\text{-}4)$$

$$[Q^-] = \frac{R(K/H^+)}{(1 + K/H^+)} \quad (20\text{-}5)$$

$$-\frac{d[R]}{dt} = \frac{2k_1 R^2}{(1 + K/H^+)^2} + \frac{2k_2 R^2 (K/H^+)}{(1 + K/H^+)^2} + \frac{2k_3 R^2 (K/H^+)^2}{(1 + K/H^+)^2} \quad (20\text{-}6)$$

$$= 2R^2 \left[\frac{k_1 + k_2(K/H^+) + k_3(K/H^+)^2}{(1 + K/H^+)^2} \right] \quad (20\text{-}7)$$

Hence the experimentally observed rate constant at a given acidity is

$$k_{obs} = \left[\frac{k_1 + k_2(K/H^+) + k_3(K/H^+)^2}{(1 + K/H^+)^2} \right] \quad (20\text{-}8)$$

REFERENCES

1. B. H. J. Bielski and E. Saito, *J. Phys. Chem.*, **66**, 2266 (1962).
2. G. Czapski and B. H. J. Bielski, *J. Phys. Chem.*, **67**, 2180 (1963).
3. B. H. J. Bielski and A. O. Allen, *Proc. 2nd Tihany Symp. Radiation Chemistry*, Akad. Kiodo, Budapest, 1967.
4. G. Czapski and L. M. Dorfman, *J. Phys. Chem.*, **68**, 1169 (1964).
5. B. H. J. Bielski and H. A. Schwarz, *J. Phys. Chem.*, **72**, 3836 (1968).

System 20. Catalyzed Decay of a Dissociable Free Radical

6. J. Rabani and S. O. Nielsen, *J. Phys. Chem.*, **73**, 3736 (1969).
7. D. Behar, G. Czapski, L. M. Dorfman, J. Rabani, and H. A. Schwarz, *J. Phys. Chem.*, **74**, 3209 (1970).
8. B. H. J. Bielski, D. A. Comstock, and R. A. Bowen, *J. Amer. Chem. Soc.* **93**, 5624 (1971).
9. B. H. J. Bielski and E. Saito, *J. Phys. Chem.*, **75**, 2263 (1971).

11

Parallel Reactions

SYSTEM 21. PARALLEL FIRST-ORDER REACTIONS WITH COMMON PRODUCT

$$A \xrightarrow{k_1} C \qquad (I)$$

$$B \xrightarrow{k_2} C \qquad (II)$$

The first-order disappearance of reactants A and B is described by the exponential equations (see p. 10)

$$[A] = [A]_0 e^{-k_1 t} \qquad (21\text{-}1)$$

$$[B] = [B]_0 e^{-k_2 t} \qquad (21\text{-}2)$$

where A, B, and $[A]_0$, $[B]_0$ are the corresponding concentrations at time $t = t$ and $t = 0$.

Since both A and B yield the same product, the rate of formation of C is given by

$$\frac{d[C]}{dt} = k_1[A] + k_2[B] \qquad (21\text{-}3)$$

System 22. Parallel First-order Reactions

Substitution of equations 21-1 and 21-2 into 21-3 yields

$$\frac{d[C]}{dt} = k_1[A]_0 e^{-k_1 t} + k_2[B]_0 e^{-k_2 t} \qquad (21\text{-}4)$$

$$\int_{[C]_0}^{[C]} d[C] = k_1[A]_0 \int_0^t e^{-k_1 t}\, dt + k_2[B]_0 \int_0^t e^{-k_2 t}\, dt \qquad (21\text{-}5)$$

$$[C] - [C]_0 = -\left\{\frac{k_1[A]_0}{k_1}\right\} e^{-k_1 t}\bigg|_0^t$$

$$\qquad\qquad - \left\{\frac{k_2[B]_0}{k_2}\right\} e^{-k_2 t}\bigg|_0^t \qquad (21\text{-}6)$$

Hence

$$[C] = [C]_0 + [A]_0 + [B]_0 - [[A]_0 e^{-k_1 t} + [B]_0 e^{-k_2 t}] \qquad (21\text{-}7)$$

Assuming that at time $t = 0$, $[C]_0 = 0$, equation 21-7 reduces to

$$[C] = ([A]_0 + [B]_0) - [[A]_0 e^{-k_1 t} + [B]_0 e^{-k_2 t}] \qquad (21\text{-}8)$$

where $[A]_0 + [B]_0 = [C]_\infty$ represents the product concentration at time $t = \infty$.

REFERENCE

1. A. A. Frost, G. G. Pearson, *Kinetics and Mechanism*, John Wiley & Sons, Inc., New York, 1961.

SYSTEM 22. PARALLEL FIRST-ORDER REACTIONS

$$A \xrightarrow{k_1} B \qquad (\text{I})$$

$$A \xrightarrow{k_2} C \qquad (\text{II})$$

$$A \xrightarrow{k_3} D \qquad (\text{III})$$

70 Parallel Reactions

This mechanism describes a system in which a single compound (*A*) changes by first-order kinetics to three different products *B*, *C*, and *D* simultaneously. At time $t = 0$ the initial concentrations of the four components will be represented by $[A]_0$, $[B]_0$, $[C]_0$, and $[D]_0$.

As is apparent, the overall rate of disappearance of *A* is the sum of all three reactions (I), (II), and (III), and can be described by the following differential equation:

$$-\frac{d[A]}{dt} = k_1[A] + k_2[A] + k_3[A] \tag{22-1}$$

$$= (k_1 + k_2 + k_3)[A] \tag{22-2}$$

$$= k[A] \tag{22-3}$$

where $k = k_1 + k_2 + k_3$.

Integration of equation 22-3 yields

$$-\int_{[A]_0}^{[A]} \frac{d[A]}{[A]} = k\int_0^t dt \tag{22-4}$$

$$-[\ln [A] - \ln [A]_0] = k(t - 0) = kt \tag{22-5}$$

$$\ln\left(\frac{[A]}{[A]_0}\right) = -kt \tag{22-6}$$

$$[A] = [A]_0 e^{-kt} \tag{22-7}$$

$$[A] = [A]_0 e^{-(k_1+k_2+k_3)t} \tag{22-8}$$

The rate of formation of *B* is given by

$$\frac{d[B]}{dt} = k_1[A] = k_1[A]_0 e^{-kt} \tag{22-9}$$

which integrates as follows:

$$\int_{[B]_0}^{[B]} d[B] = k_1[A]_0 \int_0^t e^{-kt}\, dt \tag{22-10}$$

$$[B] - [B]_0 = k_1[A]_0 \left[\frac{e^{-k \cdot 0}}{k} - \frac{e^{-kt}}{k}\right] \tag{22-11}$$

hence

$$[B] = [B]_0 + \frac{k_1[A]_0}{k}[1 - e^{-kt}] \qquad (22\text{-}12)$$

Similarly, expressions can be derived for the products C and D:

$$[C] = [C]_0 + \frac{k_2[A]_0}{k}[1 - e^{-kt}] \qquad (22\text{-}13)$$

$$[D] = [D]_0 + \frac{k_3[A]_0}{k}[1 - e^{-kt}] \qquad (22\text{-}14)$$

If at time $t = 0$, $[B]_0 = [C]_0 = [D]_0 = 0$, equations 22-12, 22-13, and 22-14 reduce to

$$[B] = \frac{k_1[A]_0}{k}[1 - e^{-kt}] \qquad (22\text{-}15)$$

$$[C] = \frac{k_2[A]_0}{k}[1 - e^{-kt}] \qquad (22\text{-}16)$$

$$[D] = \frac{k_3[A]_0}{k}[1 - e^{-kt}] \qquad (22\text{-}17)$$

indicating that $[B]:[C]:[D] = k_1:k_2:k_3$. This indicates that the concentrations of products are in a constant ratio, independent of initial concentration and time. It follows that relative rate constants can be determined by measuring the relative concentrations of the products.

REFERENCES

1. A. W. Francis, *J. Am. Chem. Soc.*, **48**, 655 (1926).
2. D. F. Smith, *J. Am. Chem. Soc.*, **49**, 43 (1927).
3. H. von Halban and H. Eisner, *Helv. Chim. Acta*, **19**, 915 (1936).
4. R. E. Fuguitt and J. E. Hawkins, *J. Am. Chem. Soc.*, **67**, 242 (1945).
5. R. E. Fuguitt and J. E. Hawkins, *J. Am. Chem. Soc.*, **69**, 319 (1947).
6. A. A. Frost and R. G. Pearson, *Kinetics and Mechanism*, John Wiley & Sons, Inc., New York, 1961.

SYSTEM 23. PARALLEL PSEUDO FIRST-ORDER REACTIONS

$$A + C \xrightarrow{k_1} \text{products} \qquad (I)$$

$$B + C \xrightarrow{k_2} \text{products} \qquad (II)$$

This mechanism consists of two reactions in which A and B compete for C. From an experimental point of view, each reaction could obviously be studied independently of the other. Hence the determination of the second order rate constants k_1 and k_2 does not constitute a particular problem.

To find a solution for the two concurrent reactions, experimental conditions can often be adjusted so that the second-order rate constants are studied under pseudo first-order conditions.

The rate expressions which describe the mechanism above are

$$-\frac{d[A]}{dt} = k_1[A][C] \qquad (23\text{-}1)$$

$$-\frac{d[B]}{dt} = k_2[B][C] \qquad (23\text{-}2)$$

Pseudo first-order conditions can be imposed by making $[C]$ much greater than $[A]$ or $[B]$. Then $[C]$ remains approximately constant during the reaction. Substitution of k_1' for $k_1[C]$ and k_2' for $k_2[C]$ into equation 23-1 and 23-2 yields

$$-\frac{d[A]}{dt} = k_1'[A] \qquad (23\text{-}3)$$

$$-\frac{d[B]}{dt} = k_2'[B] \qquad (23\text{-}4)$$

Integration yields the familiar first-order equations

$$[A] = [A]_0 e^{-k'_1 t} \tag{23-5}$$

$$[B] = [B]_0 e^{-k'_2 t} \tag{23-6}$$

from which the amounts of $[A]$ and $[B]$ which have reacted with $[C]$ in time t are given by $[A]_0[1 - e^{-k'_1 t}]$ and $[B]_0[1 - e^{-k'_2 t}]$, respectively.

The amount of $[C]$ which has reacted in time t is equal to the sum of $[A]$ and $[B]$ which have reacted in the same time interval, or

$$[C]_0 - [C] = [A]_0[1 - e^{-k'_1 t}] + [B]_0[1 - e^{-k'_2 t}] \tag{23-7}$$

REFERENCES

1. T. S. Lee, *Anal. Chem.*, **21**, 537 (1949).
2. T. S. Lee and I. M. Kolthoff, *Ann. N.Y. Acad. Sci.*, **53**, 1093 (1951).
3. N. V. Riggs, *Australian J. Chem.*, **11**, 86 (1958).
4. I. M. Kolesnikov, *Z. Fiz. Khim.*, **34**, 1069 (1960).
5. H. R. Sievert, P. N. Tenney, and T. Xermeulen, U.S. Atomic Energy Commission, UCRL 10575, 1962.

SYSTEM 24. PARALLEL REVERSIBLE AND IRREVERSIBLE REACTIONS OF FIRST-ORDER

$$A \underset{k_2}{\overset{k_1}{\rightleftharpoons}} B \tag{I}$$

$$A \xrightarrow{k_3} C \tag{II}$$

The most straightforward solution for the differential equations 24-1, 24-2, 24-3, which describes this mechanism is obtained by the application of Laplacian transforms. This method requires conditions where at time $t = 0$, $[A] = [A]_0$, and $[B]_0 = [C]_0 = 0$.

74 Parallel Reactions

The differential equations for the change in concentration with time of the individual substances are

$$\frac{d[A]}{dt} = k_2[B] - (k_1 + k_3)[A] \tag{24-1}$$

$$\frac{d[B]}{dt} = k_1[A] - k_2[B] \tag{24-2}$$

$$\frac{d[C]}{dt} = k_3[A] \tag{24-3}$$

The transformed functions (See Appendix II) for (24-1), (24-2), and (24-3) are

$$S[A] - S[A]_0 = k_2[B] - [A](k_1 + k_3) \tag{24-4}$$

$$S[B] = k_1[A] - k_2[B] \tag{24-5}$$

$$S[C] = k_3[A] \tag{24-6}$$

The expression for

$$[B] = \frac{k_1[A]}{(S + k_2)} \tag{24-7}$$

is obtained from (24-5) and substituted into (24-4):

$$S[A] - S[A]_0 = \frac{k_2 k_1 [A]}{(S + k_2)} - k_1[A] - k_3[A] \tag{24-8}$$

or

$$[A] = \left[\frac{S[A]_0 (S + k_2)}{S^2 + S(k_1 + k_2 + k_3) + k_2 k_3} \right] \tag{24-9}$$

The denominator of (24-9) can be converted by factoring to $(S + \gamma_1)(S + \gamma_2)$ if one assumes that the roots of the quadratic equation

$$S^2 + S(k_1 + k_2 + k_3) + k_2 k_3 = 0 \tag{24-10}$$

are $-\gamma_1$ and $-\gamma_2$. Hence equation 24-9 can be rewritten as

$$[A] = \frac{S[A]_0 (S + k_2)}{(S + \gamma_1)(S + \gamma_2)} \tag{24-11}$$

System 25. Parallel Second-order Reactions

Substitution for [A] in equations 24-5 and 24-6 leads to expressions for [B] and [C]:

$$[B] = \frac{k_1 S[A]_0}{(S + \gamma_1)(S + \gamma_2)} \quad (24\text{-}12)$$

$$[C] = \frac{k_3[A]_0(S + k_2)}{(S + \gamma_1)(S + \gamma_2)} \quad (24\text{-}13)$$

The final solutions are obtained by replacing the transforms with the appropriate originals from Table I, Appendix II:

$$[A] = [A]_0 \left[\left(\frac{k_2 - \gamma_1}{\gamma_2 - \gamma_1}\right) e^{-\gamma_1 t} + \left(\frac{k_2 - \gamma_2}{\gamma_1 - \gamma_2}\right) e^{-\gamma_2 t} \right] \quad (24\text{-}14)$$

$$[B] = k_1[A]_0 \left[\left(\frac{1}{\gamma_2 - \gamma_1}\right) e^{-\gamma_1 t} + \left(\frac{1}{\gamma_1 - \gamma_2}\right) e^{-\gamma_2 t} \right] \quad (24\text{-}15)$$

$$[C] = [A]_0 \left[1 - \left(\frac{k_3(k_2 - \gamma_1)}{\gamma_1(\gamma_2 - \gamma_1)}\right) e^{-\gamma_1 t} - \left(\frac{k_3(k_2 - \gamma_2)}{\gamma_2(\gamma_1 - \gamma_2)}\right) e^{-\gamma_2 t} \right]$$

$$(24\text{-}16)$$

As is apparent the roots $-\gamma_1$ and $-\gamma_2$ can be obtained in terms of k_1, k_2, and k_3 by solving equation 24-10. The derived values can then be substituted into the equations above.

REFERENCES

1. S. W. Benson, *J. Chem. Phys.*, **20**, 1605 (1952).
2. D. H. McDaniel and C. R. Smoot, *J. Phys. Chem.*, **60**, 966 (1956).
3. H. R. Sievert, P. N. Tenney, and T. Xermeulen, U.S. Atomic Energy Commission, UCRL 10575, 1962.

SYSTEM 25. PARALLEL SECOND-ORDER REACTIONS

$$A + A \xrightarrow{k_1} \text{products} \quad \text{(I)}$$

$$A + B \xrightarrow{k_2} \text{products} \quad \text{(II)}$$

$$B + B \xrightarrow{k_3} \text{products} \quad \text{(III)}$$

76 Parallel Reactions

Although no mathematical solution exists for such a mechanism when the rate constants for the constituent reactions are all different, it was shown by Hagemann and Schwarz that where certain assumptions can be justified, a reasonable interpretation of experimental results can be obtained.

By assuming that the rate constants for reactions I and III are equal to half the magnitude of the rate constant for reaction II,

$$2k_1 = k_2 = 2k_3$$

the differential equations for this system can be solved.

The rates of disappearance of reactants A and B can be expressed by the following equations:

$$\frac{d[A]}{dt} = -2k_1[A]^2 - 2k_1[A][B] \qquad (25\text{-}1)$$

$$\frac{d[B]}{dt} = -2k_1[A][B] - 2k_1[B]^2 \qquad (25\text{-}2)$$

Addition of these yields

$$\left[\frac{d[A] + d[B]}{dt}\right] = -2k_1[A]^2 - 4k_1[A][B] - 2k_1[B]^2 \qquad (25\text{-}3)$$

Inspection of the right-hand side of this equation shows that it can be expressed as binomial, a form in which it can be integrated. As is apparent, the binomial can only be obtained if the earlier assumptions concerning relative magnitudes of rate constants can be justified. Then

$$\frac{d([A] + [B])}{dt} = -2k_1([A] + [B])^2 \qquad (25\text{-}4)$$

or

$$\frac{d([A] + [B])}{([A] + [B])} = -2k_1([A] + [B])\, dt \qquad (25\text{-}5)$$

Since

$$\frac{d([A] + [B])}{([A] + [B])} = d \ln ([A] + [B]) \qquad (25\text{-}6)$$

System 25. Parallel Second-order Reactions

equation 25-5 becomes

$$\frac{d \ln ([A] + [B])}{([A] + [B])} = -2k_1 \, dt \qquad (25\text{-}7)$$

$$\int_{([A]_0+[B]_0)}^{([A]+[B])} \frac{d \ln ([A] + [B])}{([A] + [B])} = -2k_1 \int_0^t dt \qquad (25\text{-}8)$$

Hence

$$\left(\frac{1}{[A]_0 + [B]_0}\right) = \left(\frac{1}{[A] + [B]}\right) - 2k_1 t \qquad (25\text{-}9)$$

and

$$[B] = \left(\frac{[A]_0 + [B]_0}{1 + 2k_1([A]_0 + [B]_0)t}\right) - [A] \qquad (25\text{-}10)$$

By substituting the expression found for $[B]$ into equation 25-1 one obtains

$$\frac{d[A]}{dt} = -2k_1[A]\left[\frac{[A]_0 + [B]_0}{[1 + 2k_1([A]_0 + [B]_0)t]} - [A]\right] - 2k_1[A]^2 \qquad (25\text{-}11)$$

When (25-11) is divided by $[A]$ and multiplied by dt, it gives upon rearrangement

$$\int_{[A_0]}^{[A]} \frac{d[A]}{[A]} = -\int_0^t \frac{d[1 + 2k_1([A]_0 + [B]_0)t]}{[1 + 2k_1([A]_0 + [B]_0)t]} \qquad (25\text{-}12)$$

$$\ln [A] - \ln [A]_0 = -\ln [1 + 2k_1([A]_0 + [B]_0)t] \qquad (25\text{-}13)$$

$$\ln \frac{[A]_0}{[A]} = \ln [1 + 2k_1([A]_0 + [B]_0)t] \qquad (25\text{-}14)$$

hence

$$\frac{[A]_0}{[A]} = 1 + 2k_1([A]_0 + [B]_0)t \qquad (25\text{-}15)$$

or

$$k_1 = \left[\frac{[A]_0 - [A]}{2t[A]([A]_0 + [B]_0)}\right] \qquad (25\text{-}16)$$

78 Parallel Reactions

An alternate approximation of k_1 can be obtained under experimental conditions ($[B] \ll [A]$), where the second right-hand term of equation 25-2 can be neglected,

$$\frac{d[B]}{dt} = -2k_1[A][B] \qquad (25\text{-}17)$$

Substituting for A the well-known expression $[A] = [A]_0/(1 + 2k_1t[A]_0)$, where $[A]_0$ is the concentration of A at time $t = 0$, yields

$$\frac{d[B]}{dt} = -2k_1[B]\left[\frac{[A]_0}{1 + 2k_1t[A]_0}\right] \qquad (25\text{-}18)$$

$$\frac{d[B]}{[B]} = -\left[\frac{2k_1[A]_0\,dt}{1 + 2k_1t[A]_0}\right] \qquad (25\text{-}19)$$

$$\frac{d[B]}{[B]} = \frac{d(1 + 2k_1t[A]_0)}{(1 + 2k_1t[A]_0)} \qquad (25\text{-}20)$$

$$\ln\frac{[B]}{[B]_0} = -\ln(1 + 2k_1t[A]_0) \qquad (25\text{-}21)$$

$$\frac{[B]}{[B]_0} = \frac{1}{1 + 2k_1t[A]_0} \qquad (25\text{-}22)$$

Hence

$$k_1 = \left[\frac{1}{[B]} - \frac{1}{[B]_0}\right]\frac{1}{2t[A]_0}\frac{[B]_0}{} \qquad (25\text{-}23)$$

REFERENCE

1. R. J. Hagemann and H. A. Schwarz, *J. Phys. Chem.*, **71**, 2694 (1967).

SYSTEM 26. PARALLEL REACTIONS OF SECOND-ORDER

$$A + B \xrightarrow{k_1} C \qquad (\text{I})$$

$$A + B \xrightarrow{k_2} D \qquad (\text{II})$$

$$A + B \xrightarrow{k_3} E \qquad (\text{III})$$

System 26. Parallel Reactions of Second-order

To simplify the mathematical operations of such a complex system, it is assumed that the initial concentrations of A and B are the same. Hence the rate expression for the disappearance of either component is given by the following equation:

$$-\frac{d[A]}{dt} = 2k_1[A][B] + 2k_2[A][B] + 2k_3[A][B] \qquad (26\text{-}1)$$

$$= 2k[A][B] \qquad (26\text{-}2)$$

$$= 2k[A]^2 \qquad (26\text{-}3)$$

where $k = k_1 + k_2 + k_3$ is the effective rate constant for the overall complex reaction.

Integration of equation 26-3 gives the well-known second-order equation

$$[A] = \frac{[A]_0}{[1 + 2k[A]_0 t]} \qquad (26\text{-}4)$$

The concentrations of the products $[C]$, $[D]$, and $[E]$ can be computed as a function of time, $[A]_0$, k, k_1, k_2, and k_3. Since the mathematical operations are identical for all three products, $[C]$ will serve as an example:

$$\frac{d[C]}{dt} = 2k_1[A][B] = 2k_1[A]^2 \qquad (26\text{-}5)$$

Substituting the second-order equation 26-4 for $[A]$ gives

$$\frac{d[C]}{dt} = 2k_1 \frac{[A]_0^2}{[1 + 2k[A]_0 t]^2} \qquad (26\text{-}6)$$

$$\int_{[C]_0}^{[C]} d[C] = \frac{k_1[A]_0}{k} \int_{t=0}^{t=t} \frac{2k[A]_0 \, dt}{[1 + 2k[A]_0 t]^2} \qquad (26\text{-}7)$$

$$([C] - [C]_0) = \frac{k_1[A]_0}{k} \int_{t=0}^{t=t} \frac{d[1 + 2k[A]_0 t]}{[1 + 2k[A]_0 t]^2} \qquad (26\text{-}8)$$

$$([C] - [C]_0) = \frac{k_1[A]_0}{k} \left[1 - \frac{1}{[1 + 2k[A]_0 t]} \right] \qquad (26\text{-}9)$$

$$[C] = [C]_0 + \left[\frac{2[A]_0^2 k_1 t}{[1 + 2k[A]_0 t]} \right] \qquad (26\text{-}10)$$

Parallel Reactions

Similarly

$$[D] = [D]_0 + \left[\frac{2[A]_0^2 k_2 t}{1 + 2k[A]_0 t}\right] \quad (26\text{-}11)$$

$$[E] = [E]_0 + \left[\frac{2[A]_0^2 k_3 t}{1 + 2k[A]_0 t}\right] \quad (26\text{-}12)$$

If at $t = 0$, $[C]_0 = 0$, $[D]_0 = 0$, and $[E]_0 = 0$, equations 26-10, 26-11, and 26-12 reduce to

$$[C] = \left[\frac{2[A]_0^2 k_1 t}{1 + 2k[A]_0 t}\right] \quad (26\text{-}13)$$

$$[D] = \left[\frac{2[A]_0^2 k_2 t}{1 + 2k[A]_0 t}\right] \quad (26\text{-}14)$$

$$[E] = \left[\frac{2[A]_0^2 k_3 t}{1 + 2k[A]_0 t}\right] \quad (26\text{-}15)$$

REFERENCE

1. A. A. Frost and R. G. Pearson, *Kinetics and Mechanism* John Wiley & Sons, Inc., New York, 1961.

SYSTEM 27. PARALLEL REACTIONS OF SECOND-ORDER

$$A \xrightarrow{k_1} B^- + C^+ \quad (\text{I})$$

$$C^+ + B^- \xrightarrow{k_2} \text{product} \quad (\text{II})$$

$$C^+ + D \xrightarrow{k_3} D^+ + \text{product} \quad (\text{III})$$

$$B^- + D^+ \xrightarrow{k_4} \text{product} \quad (\text{IV})$$

This mechanism describes a situation often encountered in pulse radiolysis where the reactants C^+ and B^- are formed during a very short pulse (50 nsec), that is, $k_1 \gg k_2, k_3$, and k_4.

System 27. Parallel Reactions of Second-order

While according to experimental observation B^- always disappears by charge neutralization (II) and (IV), C^+ may undergo in addition to charge neutralization (II), charge transfer (III).

The rate of change in concentration of the positive ions $[C^+]$ and $[D^+]$ is described by the following two equations:

$$-\frac{d[C^+]}{dt} = k_2[C^+][B^-] + k_3[C^+][D] \tag{27-1}$$

$$\frac{d[D^+]}{dt} = k_3[C^+][D] - k_4[B^-][D^+] \tag{27-2}$$

Assuming that $k_2 = k_4 = k$, the sum of equation 27-1 and 27-2 yields

$$\frac{d([C^+] + [D^+])}{dt} = -k[B^-]([C^+] + [D^+]) \tag{27-3}$$

Since the balance of charges has to be preserved at any time, the following relation holds true:

$$[C^+] + [D^+] = [B^-] \tag{27-4}$$

Hence substitution of equation 27-4 for B^- in equation 27-3 yields

$$\frac{d([C^+] + [D^+])}{dt} = -k([C^+] + [D^+])^2 \tag{27-5}$$

$$\int_{([C^+]_0+[D^+]_0)}^{([C^+]+[D^+])} \frac{d([C^+] + [D^+])}{([C^+] + [D^+])^2} = -k\int_0^t dt \tag{27-6}$$

$$\frac{1}{([C^+]_0 + [D^+]_0)} - \frac{1}{([C^+] + [D^+])} = -kt \tag{27-7}$$

Since at $t = 0$, $[D^+]_0 = 0$, equation 27-7 reduces to

$$\frac{1}{[C^+]_0} - \frac{1}{[C^+] + [D^+]} = -kt \tag{28-8}$$

or

$$([C^+] + [D^+]) = [B^-] = \left[\frac{[C^+]_0}{1 + k[C^+]_0 t}\right] \tag{27-9}$$

82   Parallel Reactions

Substitution of the found expression for $[C^+]$ and $[B^-]$ into equation 27-2 yields

$$\frac{d[D^+]}{dt} = \left[\frac{k_3[D][C^+]_0}{1 + k[C^+]_0 t}\right] - k_3[D][D^+] - \left[\frac{k[D^+][C^+]_0}{1 + k[C^+]_0 t}\right] \quad (27\text{-}10)$$

$$= \left[\frac{k_3[D][C^+]_0}{1 + k[C^+]_0 t}\right] - \left(k_3[D] + \left[\frac{k[C^+]_0}{1 + k[C^+]_0 t}\right]\right)[D^+] \quad (27\text{-}11)$$

The first derivative of D^+ with respect to time is derived by the following mathematical operations:

1. From equation 27-9, it is known that

$$[D^+] = \left[\frac{[C^+]_0}{1 + k[C^+]_0 t}\right] - [C^+] \quad (27\text{-}12)$$

$$= \left[\frac{[C^+]_0}{1 + k[C^+]_0 t}\right]\left[1 - \frac{[C^+](1 + k[C^+]_0 t)}{[C^+]_0}\right] \quad (27\text{-}13)$$

Setting

$$y = \left[1 - \frac{[C^+](1 + k[C^+]_0 t)}{[C^+]_0}\right] \quad (27\text{-}14)$$

$$[D^+] = \left[\frac{[C^+]_0}{1 + k[C^+]_0 t}\right] y \quad (27\text{-}15)$$

2. The derivative $d[D^+]/dt$ can be obtained from equation 27-15 by applying the general form of the differential of a product

$$d(uv) = u\,dv + v\,du \quad \text{where} \quad v = \frac{x}{z}$$

and

$$dv = d\left(\frac{x}{z}\right) = \frac{z\,dx - x\,dz}{z^2}$$

hence

$$d(uv) = d\left(\frac{u \cdot x}{z}\right) = u\left[\frac{z\,dx - x\,dz}{z^2}\right] + \frac{x}{z}\,du \quad (27\text{-}16)$$

System 27. Parallel Reactions of Second-order 83

3. In applying the general derivation 27-16 to equation 27-15 yields

$$\frac{d[D^+]}{dt} = \left[\frac{[C^+]_0}{1 + k[C^+]_0 t}\right]\frac{dy}{dt} + y\frac{d}{dt}\left[\frac{[C^+]_0}{1 + k[C^+]_0 t}\right] \quad (27\text{-}17)$$

where

$$\left[\frac{[C^+]_0}{1 + k[C^+]_0 t}\right] = V = \frac{x}{z} \quad (27\text{-}18)$$

hence

$$y\frac{d}{dt}\left[\frac{[C^+]_0}{1 + k[C^+]_0 t}\right]$$
$$= y\left[\frac{(1 + k[C^+]_0 t)(d[C^+]_0/dt) - [C^+]_0(d(1 + k[C^+]_0 t)/dt)}{(1 + k[C^+]_0 t)^2}\right] \quad (27\text{-}19)$$

$$= -\left[\frac{[C^+]_0^2 ky}{(1 + k[C^+]_0 t)^2}\right] \quad (27\text{-}20)$$

Substituting equation 27-20 into equation 27-17 yields

$$\frac{d[D^+]}{dt} = \left[\frac{[C^+]_0}{1 + k[C^+]_0 t}\right]\frac{dy}{dt} - \left[\frac{[C^+]_0^2 ky}{(1 + k[C^+]_0 t)^2}\right] \quad (27\text{-}21)$$

Equating equation 27-21 with equation 27-17 gives

$$\left[\frac{k_3[D][C^+]_0}{(1 + k[C^+]_0 t)}\right] - \left[k_3[D] + \frac{k[C^+]_0}{(1 + k[C^+]_0 t)}\right]\left[\frac{[C^+]_0}{(1 + k[C^+]_0 t)}\right]y$$
$$= \left[\frac{[C^+]_0}{(1 + k[C^+]_0 t)}\right]\frac{dy}{dt} - \left[\frac{k[C^+]_0^2}{(1 + k[C^+]_0 t)^2}\right]y \quad (27\text{-}22)$$

$$\frac{dy}{dt} - \left[\frac{k[C^+]_0}{(1 + k[C^+]_0 t)}\right]y = k_3[D] - \left[k_3[D] + \frac{k[C^+]_0}{(1 + k[C^+]_0 t)}\right]y$$

$$(27\text{-}23)$$

$$\frac{dy}{dt} + k_3[D]y = k_3[D] \quad (27\text{-}24)$$

Equation 27-24 is a first-order linear differential equation, for

84 Parallel Reactions

which the integration factor is

$$e^{\int k_3[D]dt} = e^{k_3[D]t} \tag{27-25}$$

hence

$$e^{k_3[D]t}\,dy + k_3[D]e^{k_3[D]t}y\,dt = k_3[D]e^{k_3[D]t}\,dt \tag{27-26}$$

$$d[e^{k_3[D]t}y] = k_3[D]e^{k_3[D]t}\,dt \tag{27-27}$$

$$\int_0^t d[e^{k_3[D]t}y] = k_3[D]\int_0^t e^{k_3[D]t}\,dt \tag{27-28}$$

$$e^{k_3[D]t}\,y = e^{k_3[D]t} - e^{k_3[D]\times 0} = e^{k_3[D]t} - 1 \tag{27-29}$$

hence

$$y = (1 - e^{-k_3[D]t}) \tag{27-30}$$

Substitution of y into equation 27-15 gives the final answer:

$$[D^+] = \left[\frac{(1 - e^{-k_3[D]t})[C^+]_0}{(1 + k[C^+]_0 t)}\right] \tag{27-31}$$

Alternate Method

In the previous section an expression 27-11 was developed which described the change in concentration of $[D^+]$ with time:

$$\frac{d[D^+]}{dt} + \left[k_3[D] + \frac{k[C^+]_0}{(1 + k[C^+]_0 t)}\right][D^+] = \left[\frac{k_3[D][C^+]_0}{(1 + k[C^+]_0 t)}\right] \tag{27-11}$$

This is a linear differential equation of first-order for which the corresponding integrating factor is

$$\exp\int\left(k_3[D] + \frac{k[C^+]_0}{(1 + k[C^+]_0 t)}\right)dt = \exp\{(k_3[D]t + \ln[1 + k[C^+]_0 t])\} \tag{27-32}$$

Since

$$e^{\ln(1 + k[C^+]_0 t)} = (1 + k[C^+]_0 t) \tag{27-33}$$

$$\exp\left[\int\left(k_3[D] + \frac{k[C^+]_0}{(1 + k[C^+]_0 t)}\right)dt\right] = (1 + k[C^+]_0 t)e^{k_3[D]t} \tag{27-34}$$

Multiplication of both sides of equation 27-11 by the integrating factor gives

$$\underbrace{(1 + k[C^+]_0 t)e^{k_3[D]t} d[D^+] + [D^+][k_3[D]}_{} \\ + (k[C^+]_0 t)/(1 + k[C^+]_0 t)](1 + k[C^+]_0 t)e^{k_3[D]t}$$

$$= \left[\frac{k_3[D][C^+]_0(1 + k[C^+]_0 t)e^{k_3[D]t}}{(1 + k[C^+]_0 t)}\right] dt = \underbrace{[C^+]_0 k_3[D]e^{k_3[D]t} \, dt}_{}$$

(27-35)

Since those parts of equation 27-35 which are marked by horizontal brackets are the exact differentials of

$$[(1 + k[C^+]_0 t)e^{k_3[D]t}][D^+] \quad \text{and} \quad e^{k_3[D]t}$$

respectively, equation 27-35 becomes

$$d[(1 + k[C^+]_0 t)e^{k_3[D]t}[D^+]] = [C^+]_0 \, d(e^{k_3[D]t}) \quad (27\text{-}36)$$

Integration gives

$$(1 + k[C^+]_0 t)e^{k_3[D]t}[D^+] = [C^+]_0 e^{k_3[D]t} \Big|_0^t = [C^+]_0(e^{k_3[D]t} - 1)$$

(27-37)

Hence

$$[D^+] = \frac{[C^+]_0(1 - e^{-k_3[D]t})}{(1 + k[C^+]_0 t)} \quad (27\text{-}38)$$

REFERENCE

1. C. Capellos and A. O. Allen, *J. Phys. Chem.*, **73**, 3264 (1969).

SYSTEM 28. PARALLEL REACTIONS OF FIRST- AND SECOND-ORDER

$$A \xrightarrow{k_1} \text{product} \quad \text{(I)}$$
$$A + A \xrightarrow{k_2} \text{product} \quad \text{(II)}$$

86   Parallel Reactions

This system can be solved in general terms since solutions exist for the corresponding differential equations. The rate of disappearance of A is described by the following equation, where $[A]_0$ is the amount of reactant present at time $t = 0$, and $([A]_0 - x)$ at time $t = t$:

$$\frac{d([A]_0 - x)}{dt} = -k_1([A]_0 - x) - 2k_2([A]_0 - x)^2 \quad (28\text{-}1)$$

$$-\frac{dx}{dt} = -k_1([A]_0 - x) - 2k_2([A]_0 - x)^2 \quad (28\text{-}2)$$

hence

$$\frac{dx}{dt} = k_1([A]_0 - x) + 2k_2([A]_0 - x)^2 \quad (28\text{-}3)$$

and

$$\frac{dx}{([A]_0 - x)[k_1 + 2k_2([A]_0 - x)]} = dt \quad (28\text{-}4)$$

For integration equation 28-4 has to be resolved into partial fractions:

$$\frac{1}{([A]_0 - x)[k_1 + 2k_2([A]_0 - x)]} = \frac{p}{([A]_0 - x)}$$

$$+ \frac{q}{k_1 + 2k_2([A]_0 - x)} \quad (28\text{-}5)$$

$$1 = p[k_1 + 2k_2([A]_0 - x)] + q([A]_0 - x) \quad (28\text{-}6)$$

$$1 = pk_1 + 2k_2 p[A]_0 + q[A]_0 + [-(2k_2 p + q)]x \quad (28\text{-}7)$$

$$1 = pk_1 + 2k_2 p[A]_0 + q[A]_0 \quad (28\text{-}8)$$

$$2k_2 p + q = 0 \quad (28\text{-}9)$$

Hence

$$p = -\frac{q}{2k_2} \quad \text{or} \quad q = -2k_2 p \quad (28\text{-}10)$$

and substitution of these expressions into (28-8) gives

$$p = \frac{1}{k_1} \tag{28-11}$$

$$q = \frac{-2k_2}{k_1} \tag{28-12}$$

Substituting the values found for the constants p and q, equation 28-4 becomes in terms of partial fractions

$$\int_0^x \frac{dx}{k_1([A]_0 - x)} - \int_0^x \frac{2k_2\, dx}{k_1[k_1 + 2k_2([A]_0 - x)]} = \int_0^t dt \tag{28-13}$$

$$-\frac{1}{k_1}\int_0^x \frac{d([A]_0 - x)}{([A]_0 - x)} + \frac{1}{k_1}\int_0^x \frac{d[2k_2([A]_0 - x) + k_1]}{[2k_2([A]_0 - x) + k_1]} = \int_0^t dt \tag{28-14}$$

$$-\ln([A]_0 - x) + \ln[A]_0 + \ln[2k_2([A]_0 - x) + k_1]$$
$$- \ln[2k_2[A]_0 + k_1] = k_1 t \tag{28-15}$$

$$\ln\left[\frac{[A]_0[k_1 + 2k_2([A]_0 - x)]}{([A]_0 - x)[k_1 + 2k_2[A]_0]}\right] = k_1 t \tag{28-16}$$

REFERENCE

1. K. B. Yerrick and M. E. Russell, *J. Phys. Chem.*, **68**, 3752 (1964).

SYSTEM 29. PARALLEL FIRST- AND SECOND-ORDER REACTIONS

$$A \xrightarrow{k_1} D + E \tag{I}$$

$$E + B \xrightarrow{k_2} C \tag{II}$$

$$A + B \xrightarrow{k_3} C + D \tag{III}$$

This system describes the hydrolysis of organic halides (A) in the presence of hydroxide ion (B), by either an S_N1 mechanism (first-order kinetics) or an S_N2 mechanism (second-order

Parallel Reactions

kinetics). Cases are known in which both mechanisms occur simultaneously. For the latter case with conditions where $k_2 \gg k_1$ or k_3 the overall rate of reaction is given by the following differential equations:

$$\frac{dx}{dt} = k_1([A]_0 - x) + k_3([A]_0 - x)([B]_0 - x) \quad (29\text{-}1)$$

$$= ([A]_0 - x)[k_1 + k_3([B]_0 - x)] \quad (29\text{-}2)$$

or

$$\left[\frac{dx}{([A]_0 - x)[k_1 + k_3([B]_0 - x)]}\right] = dt \quad (29\text{-}3)$$

Before integration equation 29-3 has to be resolved into partial fractions:

$$\frac{1}{([A]_0 - x)(k_1 + k_3[B]_0 - k_3 x)} = \frac{p}{([A]_0 - x)}$$

$$+ \frac{q}{(k_1 + k_3[B]_0 - k_3 x)} \quad (29\text{-}4)$$

$$1 = p(k_1 + k_3[B]_0 - k_3 x) + q([A]_0 - x) \quad (29\text{-}5)$$

$$1 = pk_1 + pk_3[B]_0 + q[A]_0 - (pk_3 + q)x \quad (29\text{-}6)$$

$$pk_3 + q = 0 \quad \text{or} \quad p = \frac{-q}{k_3} \quad (29\text{-}7)$$

$$pk_1 + pk_3[B]_0 + q[A]_0 = 1 \quad (29\text{-}8)$$

$$\frac{-k_1 q}{k_3} - q[B]_0 + [A]_0 q = 1 \quad (29\text{-}9)$$

$$q\left([A]_0 - \frac{k_1}{k_3} - [B]_0\right) = 1 \quad (29\text{-}10)$$

Hence

$$p = \frac{1}{k_3[([B]_0 - [A]_0) + k_1/k_3]} \quad (29\text{-}11)$$

$$q = -\frac{1}{[k_1/k_3 + [B]_0 - [A]_0]} \quad (29\text{-}12)$$

System 30. Parallel Pseudo First- and Second-order Reactions

Equation 29-3 now becomes

$$\int_0^x \frac{dx}{k_3[([B]_0 - [A]_0) + k_1/k_3]([A]_0 - x)}$$
$$- \int_0^x \frac{dx}{[([B]_0 - [A]_0) + k_1/k_3][k_1 + k_3([B]_0 - x)]} = \int_0^t dt \quad (29\text{-}13)$$

$$- \frac{1}{k_3[([B]_0 - [A]_0) + k_1/k_3]} \int_0^x \frac{d([A]_0 - x)}{([A]_0 - x)}$$
$$+ \frac{1}{k_3[([B]_0 - [A]_0) + k_1/k_3]} \int_0^x \frac{d[([B]_0 - x)k_3 + k_1]}{[([B]_0 - x)k_3 + k_1]} = \int_0^t dt$$
$$(29\text{-}14)$$

$$\frac{1}{[([B]_0 - [A]_0) + k_1/k_3]} \left[-\ln([A]_0 - x)\Big|_0^x + \ln(k_3[k_1/k_3 + ([B]_0 - x)])\Big|_0^x \right] = k_3 t \quad (29\text{-}15)$$

or

$$\frac{1}{[([B]_0 - [A]_0) + k_1/k_3]} \left[\ln\left(\frac{[A]_0}{([A]_0 - x)}\right) \frac{[([B]_0 - x) + k_1/k_3]}{[([B]_0 + k_1/k_3)]} \right]$$
$$= k_3 t \quad (29\text{-}16)$$

SYSTEM 30. PARALLEL PSEUDO FIRST- AND SECOND-ORDER REACTIONS

$$A + B \xrightarrow{k_1} C \quad (\text{I})$$

$$A + A \xrightarrow{k_2} \text{products} \quad (\text{II})$$

This system illustrates a situation often encountered in radiation chemistry, when an active radical A formed during an electron pulse, disappears rapidly by reacting with itself. While k_2 can be determined directly in the absence of any scavengers, k_1 is determined under experimental conditions where $[B] \gg [A]$ so that $[B]$ remains virtually constant throughout the reaction. Then C is formed under pseudo first-order conditions.

Parallel Reactions

The rate at which A disappears is given by

$$-\frac{d[A]}{dt} = k_1[A][B] + 2k_2[A]^2 \qquad (30\text{-}1)$$

but since $[B]$ stays constant, $k_1[A][B] = k_1'[A]$, and

$$-\frac{d[A]}{dt} = k_1'[A] + 2k_2[A]^2 \qquad (30\text{-}2)$$

Similarly, the rate of formation of C can be written

$$\frac{d[C]}{dt} = k_1[A][B] = k_1'[A] \qquad (30\text{-}3)$$

In dividing equation 30-2 by 30-3 time is eliminated:

$$-\frac{d[A]}{d[C]} = \frac{k_1'[A] + 2k_2[A]^2}{k_1'A} \qquad (30\text{-}4)$$

$$= \frac{k_1' + 2k_2[A]}{k_1'} \qquad (30\text{-}5)$$

or

$$\frac{d[C]}{d[A]} = -\frac{k_1'}{k_1' + 2k_2[A]} \qquad (30\text{-}6)$$

In order to solve equation 30-6, it has to be transformed into a form for which a mathematical solution exists

$$-\frac{k_1'}{k_1' + 2k_2[A]}\, d[A] = d[C] \qquad (30\text{-}7)$$

Multiplying the left-hand numerator and denominator by $2k_2$ yields the following integral:

$$-\frac{k_1'}{2k_2}\int \frac{2k_2\, d[A]}{(k_1' + 2k_2[A])} = \int d[C] \qquad (30\text{-}8)$$

or

$$-\int_{[A]_0}^{[A]} \frac{2k_2\, d[A]}{(k_1' + 2k_2[A])} = \frac{2k_2}{k_1'}\int_{[C]_0=0}^{[C]} d[C] \qquad (30\text{-}9)$$

System 30. Parallel Pseudo First- and Second-order Reactions

The solution for the left-hand side of equation 30-9 is the known integral

$$\int \frac{f'(x)\,dx}{f(x)} = \ln f(x) + I \tag{30-10}$$

where $f'(x)$ is the first derivative of $f(x)$. As is apparent the numerator $2k_2 d[A]$ is the first derivative of the denominator $(k_1' + 2k_2[A])$ in equation 30-9. Hence integration of 30-9 between the corresponding limits yields

$$\ln\left[\frac{(k_1' + 2k_2[A])}{(k_1' + 2k_2[A]_0)}\right] = \frac{-2k_2[C]}{k_1'} \tag{30-11}$$

Setting $2k_2/k_1' = u$, equation 30-11 can be written in the exponential form

$$\left[\frac{k_1' + 2k_2[A]}{k_1' + 2k_2[A]_0}\right] = e^{-u[C]} \tag{30-12}$$

or

$$[A] = \frac{1}{u}[e^{1/u[C]} + e^{u[A]_0/u[C]} - 1] \tag{30-13}$$

$$= \frac{1}{u}\left[\frac{1 + u[A]_0}{e^{u[C]}} - 1\right] \tag{30-14}$$

A further simplification is carried out in setting $(1 + u[A]_0) = z$:

$$[A] = \frac{1}{u}\left[\frac{z - e^{u[C]}}{e^{u[C]}}\right] \tag{30-15}$$

Substitution of the expression for $[A]$ into the original rate equation for the formation of C yields

$$\frac{d[C]}{dt} = \frac{k_1'}{u}\left[\frac{z - e^{u[C]}}{e^{u[C]}}\right] \tag{30-16}$$

Integration between concentration limits leads to an expression

Parallel Reactions

for which [C] can be determined at any time t in terms of k_2, k_1', and $[A]_0$

$$\int_{[C_0]}^{[C]} \frac{ue^{u[C]}d[C]}{(z - e^{u[C]})} = k_1' \int_0^t dt \qquad (30\text{-}17)$$

$$-\int_{[C]_0}^{[C]} \frac{d(z - e^{u[C]})}{(z - e^{u[C]})} = k_1' t \qquad (30\text{-}18)$$

$$\ln\left[\frac{z - e^{u[C]}}{z - 1}\right] = -k_1' t \qquad (30\text{-}19)$$

$$\frac{(z - e^{u[C]})}{(z - 1)} = e^{-k_1' t} \qquad (30\text{-}20)$$

or

$$[C] = \frac{1}{u}\ln\left[z - (z - 1)e^{-k_1' t}\right] \qquad (30\text{-}21)$$

REFERENCE

1. J. Rabani and S. O. Nielsen, *J. Phys. Chem.*, **73**, 3736 (1969).

12

Consecutive Reactions of Higher Order

Chemical systems that follow a sequence of consecutive higher-order reactions are in general described by nonlinear differential equations that do not have exact solutions.

There are a few exceptions to this generalization, but in the majority of cases solutions are not available, hence alternate mathematical operations must be used in studies of such systems.

A frequently used simplification is based on the elimination of time as an independent variable. This method is of use only if the original set of equations defining the chemical system are such that the elimination of time as an independent variable leads to a new set of differential equations in which the remaining variables are separable.

SYSTEM 31. FIRST-ORDER REACTION FOLLOWED BY A SECOND-ORDER REACTION

$$A \xrightarrow{k_1} B \qquad (I)$$

$$A + B \xrightarrow{k_2} C \qquad (II)$$

94 Consecutive Reactions of Higher Order

This complex mechanism is described by the following set of nonlinear differential equations:

$$\frac{d[A]}{dt} = -k_1[A] - k_2[A][B] \tag{31-1}$$

$$\frac{d[B]}{dt} = k_1[A] - k_2[A][B] \tag{31-2}$$

$$\frac{d[C]}{dt} = k_2[A][B] = -\frac{1}{2}\left\{\frac{d[A]}{dt} - \frac{d[B]}{dt}\right\} \tag{31-3}$$

Although no exact mathematical solution has so far been derived for such a mechanism, some information about a system of this kind can be obtained if the elimination of time as an independent variable leads to a new set of differential equations for which solutions exist.

In this case it is achieved by dividing equation 31-2 by 31-1 which gives

$$\frac{d[B]}{d[A]} = \frac{k_1[A] - k_2[A][B]}{-k_1[A] - k_2[A][B]} \tag{31-4}$$

$$= \frac{[B] - k}{[B] + k} \tag{31-5}$$

where $k = k_1/k_2$. Equation 31-5 can be rewritten as

$$\frac{k + [B]}{k - [B]} d[B] = -[dA] \tag{31-6}$$

Following the left-hand side of equation 31-6 can be transformed to read

$$\frac{k + [B]}{k - [B]} d[B] = -\frac{(-k - [B] + 2k - 2k)\,d[B]}{(k - [B])} \tag{31-7}$$

$$= -\frac{(k - [B])}{(k - [B])} d[B] + \frac{2k\,d[B]}{(k - [B])} \tag{31-8}$$

$$= -d[B] + \frac{2k}{(k - [B])} d[B] \tag{31-9}$$

Substitution of the right-hand side of equation 31-9 into equation 31-6 leads to a form in which the latter can be integrated over the respective limits:

$$-\int_{[B]_0}^{[B]} d[B] - 2k \int_{[B]_0}^{[B]} \frac{d(k - [B])}{(k - [B])} = -\int_{[A]_0}^{[A]} d[A] \quad (31\text{-}10)$$

$$-[B] + 0 - 2k[\ln(k - B) - \ln k] = -([A] - [A]_0)$$
$$(31\text{-}11)$$

Hence

$$[B] + 2k \ln\left[\frac{k - [B]}{k}\right] = ([A] - [A]_0) \quad (31\text{-}12)$$

$$\frac{[B]}{k} + 2\ln\left(1 - \frac{[B]}{k}\right) = \left[\frac{[A]_0}{k}\right]\left[\frac{[A]}{[A]_0} - 1\right] \quad (31\text{-}13)$$

REFERENCE

1. S. W. Benson, *J. Chem. Phys.*, **20**, 1605 (1952).

SYSTEM 32. SECOND-ORDER REACTION FOLLOWED BY A FIRST-ORDER REACTION

$$2A \xrightarrow{k_1} B \xrightarrow{k_2} C \quad \text{(I)}$$

Amount at time $t = 0$ $[A]_0$ 0   0

Amount at time $t = t$ $[A]$ $[B]$ $[C]$

The mathematical solution of the kinetics of this system requires the assumption that the following relationship holds true at any time,

$$[A]_0 = [A] + [B] + [C] \quad (32\text{-}1)$$

The change in concentration of the individual species is given

96 Consecutive Reactions of Higher Order

by the following set of equations:

$$-\frac{d[A]}{dt} = 2k_1[A]^2 \tag{32-2}$$

$$\frac{d[C]}{dt} = k_2[B] \tag{32-3}$$

and the rate of change in the concentration of [B] is the difference between (33-2) and (33-3)

$$\frac{d[B]}{dt} = -\frac{d[A]}{dt} - \frac{dC}{dt} = 2k_1[A]^2 - k_2[B] \tag{32-4}$$

or

$$\frac{d[B]}{dt} + k_2[B] = 2k_1[A]^2 \tag{32-5}$$

Substitution of the well-known rate expression $[A] = [A]_0/(1 + [A]_0 k_1 t)$ into equation 32-5 yields a linear differential equation of first-order

$$\frac{d[B]}{dt} + k_2[B] = \left[\frac{2k_1[A]_0^2}{(1 + 2[A]_0 k_1 t)^2}\right] \tag{32-6}$$

which is of the general form

$$\frac{dy}{dt} + P(x)y = Q(x) \tag{32-7}$$

The integrating factor for (32-6) is

$$\mu(x) = e^{\int P(x)dx} = e^{k_2 t} \tag{32-8}$$

Multiplication of equation 32-6 by the factor 32-8 and rearrangement of the terms yields

$$e^{k_2 t}d[B] + k_2 e^{k_2 t}[B]\,dt = 2k_1 e^{k_2 t}\left[\frac{[A]_0^2}{(1 + 2k_1[A]_0 t)^2}\right]dt \tag{32-9}$$

$$d[e^{k_2 t}[B]] = 2k_1 e^{k_2 t}\left[\frac{[A]_0^2}{(1 + 2k_1[A]_0 t)^2}\right]dt \tag{32-10}$$

System 32. First-order Reaction 97

Although the integration of equation 32-10 poses no serious problem, several steps are required for the solution of the right-hand side:

$$\int_0^{[B]} d[e^{k_2 t}[B]] = 2k_1[A]_0^2 \int_0^t \frac{e^{k_2 t}}{(1 + 2k_1[A]_0 t)^2} dt \quad (32\text{-}11)$$

$$e^{k_2 t}[B] = [A]_0 \int_0^t \frac{2e^{k_2 t}[A]_0 k_1}{(1 + 2k_1[A]_0 t)^2} dt \quad (32\text{-}12)$$

$$= [A]_0 \int_0^t \frac{e^{k_2 t} d(1 + 2k_1[A]_0 t)}{(1 + 2k_1[A]_0 t)^2} \quad (32\text{-}13)$$

Equation 32-13 has to be simplified by setting $(1 + 2k_1[A]_0 t) = w$ so that

$$t = \left[\frac{(w-1)}{2[A]_0 k_1}\right] \quad (32\text{-}14)$$

and

$$\frac{dw}{dt} = \frac{d(1 + 2k_1[A]_0 t)}{dt} = 2[A]_0 k_1 \quad (32\text{-}15)$$

and

$$dt = \frac{dw}{2k_1[A_0]} \quad (32\text{-}16)$$

Substitution of these terms into equation 32-13 yields

$$e^{k_2 t}[B] = [A]_0 \int_{w=1}^{w=(1+2k_1[A]_0 t)} \left[\frac{e^{[k_2(w-1/2[A]_0 k_1)]}}{w^2}\right] dw \quad (32\text{-}17)$$

$$= [A]_0 \int_1^{(1+k_1[A]_0 t)} \left[\frac{e^{(k_2 w/2k_1[A]_0) - (k_2/2k_1[A]_0)}}{w^2}\right] dw \quad (32\text{-}18)$$

$$= [A]_0 e^{-(k_2/2k_1[A]_0)} \int_1^{(1+2k_1[A]_0 t)} \left[\frac{e^{(k_2 w/2k_1[A]_0)}}{w^2}\right] dw \quad (32\text{-}19)$$

Equation 32-19 can be further simplified by setting $k_2/(2k_1[A]_0) = a$

$$e^{k_2 t}[B] = [A]_0 e^{-a} \int_1^{(1+2k_1[A]_0 t)} \frac{e^{aw} dw}{w^2} \quad (32\text{-}20)$$

98 Consecutive Reactions of Higher Order

The integral in this equation is of the general form

$$\int \frac{e^{aw} \, dw}{w^m} = \int e^{aw} w^{-m} \, dw \qquad (32\text{-}21)$$

where in this specific case $m = 2$. Since

$$w^{-m} \, dw = \frac{-(m-1) w^{-m} \, dw}{-(m-1)} = -\frac{1}{(m-1)} d[w^{-(m-1)}] \qquad (32\text{-}22)$$

the integral becomes

$$\int \frac{e^{aw}}{w^m} \, dw = -\frac{1}{(m-1)} \int e^{aw} \, d[w^{-(m-1)}] \qquad (32\text{-}23)$$

The two variable terms on the right-hand side are integrated according to rule for products such as $u \, dv$, where $\int u \, dv = uv - \int v \, du$. Equation 32-23 is now written

$$\int \frac{e^{aw}}{w^m} \, dw = -\frac{1}{(m-1)} \int e^{aw} \, d[w^{-(m-1)}] \qquad (32\text{-}24)$$

$$= -\frac{e^{aw}}{(m-1) w^{(m-1)}} + \frac{1}{(m-1)} \int \frac{d(e^{aw})}{w^{(m-1)}} \qquad (32\text{-}25)$$

$$\frac{e^{k_2 t [B]}}{[A]_0 e^{-a}} + \frac{e^{aw}}{(m-1) w^{(m-1)}} = \frac{1}{(m-1)} \int \frac{d(e^{aw})}{w^{(m-1)}} \qquad (32\text{-}26)$$

$$= \frac{a}{(m-1)} \int \frac{e^{aw} \, dw}{w} \qquad (32\text{-}27)$$

Equation 32-27 can be integrated by the use of series

$$e^u = 1 + u + \frac{u^2}{2!} + \frac{u^3}{3!} + \cdots + \frac{u^{\mu-1}}{(\mu-1)!} \qquad (32\text{-}28)$$

If $u = aw$,

$$e^{aw} = 1 + aw + \frac{a^2 w^2}{2!} + \frac{a^3 w^3}{3!} + \cdots \qquad (32\text{-}29)$$

System 32. First-order Reaction

and substitution of equation 32-29 into 32-27 yields

$$\frac{e^{k_2 t[B]}}{[A]_0 e^{-a}} + \frac{e^{aw}}{(m-1)w^{(m-1)}} = \frac{a}{(m-1)} \int \left(1 + aw + \frac{a^2 w^2}{2!}\right.$$

$$\left. + \frac{a^3 w^3}{3!} + \cdots \right) \frac{dw}{w} \quad (32\text{-}30)$$

$$= \frac{a}{(m-1)} \left[\log w + \frac{aw}{1!} + \frac{a^2 w^2}{2 \cdot 2!} + \frac{a^3 w^3}{3 \cdot 3!} + \cdots\right]_1^w$$

$$(32\text{-}31)$$

$$= \left[-\frac{a^{aw}}{w}\right]_1^w + a\left(\log w + \frac{aw}{1!} + \frac{a^2 w^2}{2 \cdot 2!} + \frac{a^3 w^3}{3 \cdot 3!}\right.$$

$$\left. - \log 1 - \frac{a}{1!} - \frac{a^2}{2 \cdot 2!} - \frac{a^3}{3 \cdot 3!} - \cdots\right] \quad (32\text{-}32)$$

$$= e^a - \frac{e^{aw}}{w} + a\left[\log w + \frac{a}{1!}(w-1) + \frac{a^2}{2 \cdot 2!}(w^2-1)\right.$$

$$\left. + \frac{a^3}{3 \cdot 3!}(w^3-1) + \cdots\right] \quad (32\text{-}33)$$

Therefore

$$e^{k_2 t}[B] = [A]_0 e^{-a}\left[e^a - \frac{e^{aw}}{w} + a\left\{\log w + \frac{a}{1!}(w-1)\right.\right.$$

$$\left.\left. + \frac{a^2}{2 \cdot 2!}(w^2-1) + \frac{a^3}{3 \cdot 3!}(w^3-1) + \cdots\right\}\right] \quad (32\text{-}34)$$

From the expressions $a = k_2/([A]_0 k_1)$ and $w = 1 + 2[A]_0 k_1 t$ it is apparent that $k_2 t = a(w-1)$, hence

$$[B] = \frac{[A]_0 e^{-a}}{e^{a(w-1)}}\left[e^a - \frac{e^{aw}}{w} + a\left\{\log w + \frac{a}{1!}(w-1)\right.\right.$$

$$\left.\left. + \frac{a^2}{2 \cdot 2!}(w^2-1) + \frac{a^3}{3 \cdot 3!}(w^3-1) + \cdots\right\}\right] \quad (32\text{-}35)$$

From the equation of material balance 32-1, $[C] = [A]_0 - [A] - [B]$, hence

$$[C] = [A]_0 - \frac{[A]_0}{w} - [A]_0 e^{aw}\left[e^a - \frac{e^{aw}}{w} + a\left(\log w + \frac{a}{1!}(w-1)\right.\right.$$

$$\left.\left. + \frac{a^2}{2 \cdot 2!}(w^2 - 1) + \frac{a^3}{3 \cdot 3!}(w^3 - 1) + \cdots\right)\right] \quad (32\text{-}36)$$

SYSTEM 33. SECOND-ORDER REACTION FOLLOWED BY A SECOND-ORDER REACTION

$$A + B \xrightarrow{k_1} C \quad \text{(I)}$$

$$B + C \xrightarrow{k_2} D \quad \text{(II)}$$

The change in concentration with time for the individual components of this system is given by the following rate expressions:

$$\frac{d[A]}{dt} = -k_1[A][B] \quad (33\text{-}1)$$

$$\frac{d[B]}{dt} = -k_1[A][B] - k_2[B][C] \quad (33\text{-}2)$$

$$\frac{d[C]}{dt} = k_1[A][B] - k_2[B][C] \quad (33\text{-}3)$$

$$\frac{d[D]}{dt} = k_2[B][C] \quad (33\text{-}4)$$

Since these equations cannot be solved by a routine mathematical operation, the method of elimination of time as an independent variable will be applied. Dividing (33-2) by (33-1) yields

$$\frac{d[B]}{d[A]} = 1 + \frac{k_2}{k_1}\frac{[C]}{[A]} = 1 + ky \quad (33\text{-}5)$$

System 33. Second-order Reaction

where $k_2/k_1 = k$ and $[C]/[A] = y$. Similarly the division of (33-3) by (33-1) yields

$$\frac{d[C]}{d[A]} = -1 + \frac{k_2}{k_1}\frac{[C]}{[A]} = -1 + ky \tag{33-6}$$

or $[C] = [A]y$, which can be differentiated:

$$d[C] = [A]\,dy + y\,d[A] \tag{33-7}$$

$$\frac{d[C]}{d[A]} = [A]\frac{dy}{d[A]} + y \tag{33-8}$$

hence

$$\frac{d[C]}{d[A]} = \frac{dy}{(1/[A])d[A]} + y = \frac{dy}{d\ln[A]} + y \tag{33-9}$$

Combining equations 33-9 and 33-6 gives

$$\frac{d[C]}{d[A]} = \frac{dy}{d\ln[A]} + y = -1 + ky \tag{33-10}$$

hence

$$\frac{dy}{d\ln[A]} = (k-1)y - 1 \tag{33-11}$$

This can be integrated

$$\int_0^y \frac{dy}{(k-1)y - 1} = \int_{[A]_0}^{[A]} d\ln[A] \tag{33-12}$$

$$\frac{1}{(k-1)}\int_0^y \frac{(k-1)\,dy}{(k-1)y - 1} = \ln[A] - \ln[A]_0 \tag{33-13}$$

$$\frac{1}{(k-1)}\int_0^y \frac{d[(k-1)y - 1]}{[(k-1)y - 1]} = \ln\frac{[A]}{[A]_0} \tag{33-14}$$

$$\frac{1}{(k-1)}[\ln[(k-1)y - 1] - \ln(-1)] = \ln\frac{[A]}{[A]_0} \tag{33-15}$$

$$\ln\left[-\left[(k-1)\frac{[C]}{[A]} - 1\right]\right] = (k-1)\ln\left[\frac{[A]}{[A]_0}\right]$$

$$= \ln\left[\frac{[A]}{[A]_0}\right]^{(k-1)} \tag{33-16}$$

102 Consecutive Reactions of Higher Order

The minus sign in front of the inner bracket arises from ln (-1). Finally, expressing the concentration of C in terms of A gives

$$\left[(k-1)\frac{[C]}{[A]}\right] - 1 = -\left[\frac{[A]}{[A]_0}\right]^{(k-1)} \quad (33\text{-}17)$$

$$[C] = \frac{[A]}{(k-1)}\left[1 - \left[\frac{[A]}{[A]_0}\right]^{(k-1)}\right] \quad (33\text{-}18)$$

To correlate $[B]$ with $[B]_0$, $[A]$ and k, the expression found for $[C]$ (equation 33-18) is substituted into equation 33-5, rearranged and integrated between the corresponding limits:

$$\frac{d[B]}{d[A]} = 1 + k\frac{[C]}{[A]} = 1 + \frac{k}{(k-1)}\left[1 - \left[\frac{[A]}{[A]_0}\right]^{(k-1)}\right] \quad (33\text{-}19)$$

$$d[B] = \left[1 + \frac{k}{(k-1)}\right]d[A] - \frac{k}{(k-1)}\left[\frac{[A]}{[A]_0}\right]^{(k-1)}d[A] \quad (33\text{-}20)$$

$$d[B] = \left[\frac{2k-1}{k-1}\right]d[A] - \frac{k}{(k-1)}\left[\frac{[A]}{[A]_0}\right]^{(k-1)}d[A] \quad (33\text{-}21)$$

$$\int_{[B]_0}^{[B]} d[B] = \int_{[A]_0}^{[A]} \frac{(2k-1)d[A]}{(k-1)} - \frac{k}{(k-1)}\int_{[A]_0}^{[A]} \frac{[A]^{(k-1)}}{[A]_0^{(k-1)}}d[A] \quad (33\text{-}22)$$

$$[B] - [B]_0 = \left[\frac{(2k-1)}{(k-1)}\right]([A] - [A]_0) - \frac{[A]_0}{(k-1)[A]_0^k}\int_{[A]_0}^{[A]} d[A]^k \quad (33\text{-}23)$$

$$[B] - [B]_0 = \left[\frac{(2k-1)}{(k-1)}\right]([A] - [A]_0)$$
$$- \left[\frac{A_0}{(k-1)[A]_0^k}\right]\left[[A]^k - [A]_0^k\right] \quad (33\text{-}24)$$

Hence

$$[B]_0 - [B] = \left[\frac{(2k-1)}{(k-1)}\right][[A]_0 - [A]] - \frac{[A]_0}{(k-1)}\left[1 - \left[\frac{[A]}{[A]_0}\right]^k\right] \quad (33\text{-}25)$$

An equation for the change in concentration of product D with time is obtained by substitution of the appropriate expressions into equation 33-26

$$[D] = [A]_0 - [A] - [C] \qquad (33\text{-}26)$$

$$[D] = [A]_0 - \frac{[A]}{(k-1)}\left[k - \left[\frac{[A]}{[A]_0}\right]^{(k-1)}\right] \qquad (33\text{-}27)$$

REFERENCES

1. G. Natta, *Rend. Inst. Lombardo Sci.*, **78,** No. 1, 307 (1945).
2. C. Potter and W. C. MacDonald, *Can. J. Res.*, **258,** 415 (1947).
3. S. W. Benson, *J. Chem. Phys.*, **20,** 1605 (1952).

13

Miscellaneous

SYSTEM 34. ELECTRON TRANSFER REACTIONS IN POLAR SOLVENTS

$$A + e \xrightarrow{k_1} B^- \qquad (I)$$

$$B^- + C \xrightarrow{k_2} C^- + B \qquad (II)$$

$$C^- + A \xrightarrow{k_3} D^- + \text{product} \qquad (III)$$

The mechanism shows three consecutive pseudo first-order reactions ($[A]$, $[C] \gg [e]$, $[B^-]$, $[C^-]$) which describe the attachment and transfer of an electron deposited into a polar solvent by pulse radiolysis. The electron, which at first attaches itself to a polar solvent molecule A, is transferred to an electron scavenger C. The product C^- can undergo further reaction with the solvent A, resulting in the formation of D^-. Since the solvent A is present in very high concentration, reaction I goes to completion very quickly and reactions II and III become rate determining.

Of interest in this special case is the rate expression for the net formation of C^-, since it is the species which is being observed in the experiment. Hence in general,

$$\frac{d[C^-]}{dt} = k_2[B^-][C] - k_3[A][C^-] \qquad (34\text{-}1)$$

System 34. Electron Transfer Reactions in Polar Solvents

but since [A], the concentration of the solvent, is effectively constant, it can be included into the pseudo first order rate constant k_3':

$$k_3[A][C^-] = k_3'[C^-] \qquad (34\text{-}2)$$

hence

$$\frac{d[C^-]}{dt} = k_2[B^-][C] - k_3'[C^-] \qquad (34\text{-}3)$$

Similarly the rate of formation of D^- is given by

$$\frac{d[D^-]}{dt} = k_3[A][C^-] = k_3'[C^-] \qquad (34\text{-}4)$$

Although there is no mathematical solution for either (34-3) or (34-4), the sum of the two equations can be solved for $[C^-]$ as a function of time:

$$\frac{d([D^-] + [C^-])}{dt} = k_2[B^-][C] \qquad (34\text{-}5)$$

This equation is transformed by use of the material balance relationship,

$$[D^-] + [C^-] = [B^-]_0 - [B^-] \qquad (34\text{-}6)$$

in which $[B^-]_0$ is the initial concentration of anion B, to

$$\frac{d([B^-]_0 - [B^-])}{dt} = -\frac{d[B^-]}{dt} = k_2[B^-][C] \qquad (34\text{-}7)$$

Since it was earlier postulated that $[C] \gg [B^-]$, the $[C]$ term can be included in the pseudo first-order rate constant, so that

$$k_2[B^-][C] = k_2'[B^-] \qquad (34\text{-}8)$$

and

$$-\frac{d[B^-]}{dt} = k_2'[B^-] \qquad (34\text{-}9)$$

On integration,

$$[B^-] = [B^-]_0 e^{-k_2't} \qquad (34\text{-}10)$$

Substitution of the expression for $[B^-]$ into equation 34-3 yields

$$\frac{d[C^-]}{dt} = k_2'[B^-]_0 e^{-k_2't} - k_3'[C^-] \tag{34-11}$$

which upon rearrangement gives the more familiar form

$$d[C^-] + k_3'[C^-]\,dt = k_2'[B^-]_0 e^{-k_2't} \tag{34-12}$$

Integration of this linear equation yields

$$[C^-] = \left(\frac{k_2'[B^-]_0}{k_2' - k_3'}\right)\left[1 - e^{-(k_2'-k_3')t}\right]e^{-k_3't} \tag{34-13}$$

REFERENCE

1. C. Capellos and A. O. Allen, *J. Phys. Chem.*, **74**, 840 (1970).

SYSTEM 35. DIFFUSION-CONTROLLED REACTIONS IN SOLUTION

A relationship that describes the rate constant of a diffusion-controlled reaction in terms of the effective collision distance and the diffusion coefficients of the interacting species can be derived in the following way.

The system consists of two interacting particles A and B in solution. Since the particles are in motion it is convenient to fix the center of coordinates on one of them, let us say A. Furthermore, the relative motion of A and B is governed by the sum of the two diffusion coefficients $(D_A + D_B)$. Hence if the total inward flow of B through a sphere of radius r about A is designated by I, then the flux of B per cm² is

$$\left[\frac{I}{4\pi r^2}\right] = (D_A + D_B)\frac{dn}{dr} \tag{35-1}$$

where n represents the concentration of species B.

Equation 35-1 can be integrated directly

$$\frac{I}{4\pi} \int_{r=R_{AB}}^{r=R_\infty} \frac{dr}{r^2} = -(D_A + D_B) \int_{n=0}^{n=n_\infty=n_B} dn \cdots \quad (35\text{-}2)$$

$$-\frac{I}{4\pi r}\bigg|_{r=R_{AB}}^{r=R_\infty} = -(D_A + D_B)n \bigg|_{n=0}^{n=n_B} \cdots \quad (35\text{-}3)$$

$$-\frac{I}{4\pi R_\infty} + \frac{I}{4\pi R_{AB}} = (D_A + D_B)n_B - (D_A + D_B) \times 0 \quad (35\text{-}4)$$

Since at $r = R_\infty$ the flux is zero, equation 35-4 reduces to

$$\frac{I}{4\pi R_{AB}} = (D_A + D_B)n_B \quad (35\text{-}5)$$

or

$$\frac{I}{n_B} = K = 4\pi R_{AB}(D_A + D_B) \cdots \quad (35\text{-}6)$$

where $I/n_B = K$ (expressed in cubic centimeters per second) represents the collision frequency of particle A with particle B, and R_{AB} is the encounter radius.

K can also be expressed in the more practical terms of 1 mole^{-1} sec^{-1},

$$K = \frac{4\pi R_{AB}(D_A + D_B)N}{10^3} \quad (35\text{-}7)$$

where N is Avogadro's number.

SYSTEM 36. EFFECT OF COULOMBIC INTERACTIONS ON REACTION VELOCITIES

This system is similar to the previous one, except that particles A and B carry charges Z_A and Z_B. As in the former case, the center of coordinates are fixed on particle A, so that the relative motion of the species will be governed by the sum of the diffusion coefficients $(D_A + D_B)$, and the coulombic potential U. If the total net inward flow of B through a sphere

108   Miscellaneous

of radius r about A is represented by I, then the flux per square centimeter is

$$\frac{I}{4\pi r^2} = +(D_A + D_B)\frac{dn}{dr} + (U_A + U_B)n\frac{dU}{dr} \quad (36\text{-}1)$$

where n represents the concentration of B, and $(U_A + U_B)$ the sum of ionic mobilities.

Equation 36-1 gives upon rearrangement

$$\frac{I}{4\pi(U_A + U_B)}\frac{dr}{r^2} = +\frac{(D_A + D_B)}{(U_A + U_B)}dn + nd(U) \quad (36\text{-}2)$$

This can be transformed into a linear differential equation by substituting the thermal energy term KT from the relationship $KT = D/U$:

$$\frac{I}{4\pi(U_A + U_B)}\frac{dr}{r^2} = +KTdn + nd(U) \quad (36\text{-}3)$$

or

$$dn + \left[\frac{dU}{KT}\right]n = \frac{I}{4\pi(D_A + D_B)}\frac{dr}{r^2} \quad (36\text{-}4)$$

To solve this equation it has to be multiplied by the integration factor

$$u = e^{\int dU/KT} = e^{(U/KT)} \quad (36\text{-}5)$$

Hence

$$e^{(U/KT)}dn + \left\{\left[\frac{1}{KT}\right]e^{(U/KT)}d(U)\right\}n = \frac{Ie^{(U/KT)}}{4\pi(D_A + D_B)}\frac{dr}{r^2} \quad (36\text{-}6)$$

or

$$d[ne^{(U/KT)}] = \frac{I}{4\pi(D_A + D_B)}(e^{(U/KT)})\frac{dr}{r^2} \quad (36\text{-}7)$$

Equation 36-7 is integrated between limits

$$n\int_{U=U_{RAB}}^{U=U_\infty=0} d(e^{(U/KT)}) = \frac{I}{4\pi(D_A + D_B)}\int_{r=R_{AB}}^{r=R\infty}(e^{(U/KT)})\frac{dr}{r^2} \quad (36\text{-}8)$$

$$n(e^{(U/KT)})\Big|_{U=U_{RAB};\ n=0}^{U=U_\infty;\ n=n_2} = n_2 = \frac{I}{4\pi(D_A + D_B)}\int_{r=R_{AB}}^{r=R\infty}(e^{(U/KT)})\frac{dr}{r^2}$$
$$(36\text{-}9)$$

System 36. Effect of Coulombic Interactions

The coulombic potential can be replaced by the relationship

$$U = \frac{[Z_A Z_B e^2]}{r\varepsilon} \tag{36-10}$$

where r is the distance between ions, e is the charge, and ε is the dielectric constant of the solvent. Hence

$$n_2 = \frac{I}{4\pi(D_A + D_B)} \int_{r=R_{AB}}^{r=R\infty} (e^{(Z_A Z_B e^2/\varepsilon rKT)}) \frac{dr}{r^2} \tag{36-11}$$

To integrate (36-11) it is expedient to replace $1/r$ by w, since it can be shown that $dr = -dw/w^2$ and $dr/r^2 = -dw$; hence

$$n_2 = -\frac{I}{4\pi(D_A + D_B)} \int_{w=1/R_{AB}}^{w=1/R\infty} e^{(Z_A Z_B e^2/\varepsilon KT)w} \, dw \tag{36-12}$$

$$n_2 = -\left(\frac{I/4\pi(D_A + D_B)}{(Z_A Z_B e^2/\varepsilon KT)}\right)(e^{(Z_A Z_B e^2/\varepsilon KT)w})\Big|_{w=1/R_{AB}}^{w=1/R\infty} \tag{36-13}$$

or

$$n_2 = \left(\frac{I/4\pi(D_A + D_B)}{[Z_A Z_B e^2/\varepsilon KT]}\right)[e^{(Z_A Z_B e^2/R_{AB}KT)} - 1] \tag{36-14}$$

hence

$$K = \frac{I}{n_2} = \frac{4\pi(D_A + D_B)(Z_A Z_B e^2/\varepsilon KT)}{[e^{(Z_A Z_B e^2/R_{AB}\varepsilon KT)} - 1]} \tag{36-15}$$

where K is expressed in cubic centimeters per second. The K expressed in $1\,\text{mole}^{-1}\text{sec}^{-1}$ gives the more familiar Debye equation

$$K = \frac{4\pi R_{AB}(D_A + D_B)[Z_A Z_B e^2/\varepsilon KTR_{AB}]N}{[e^{(Z_A Z_B e^2/R_{AB}\varepsilon KT)} - 1]10^3} \tag{36-16}$$

where N is Avogadro's number.

REFERENCES

1. P. Debye, *Trans. Electrochem. Soc.*, **82**, 265 (1942).
2. M. Eigen, W. Kruse, G. Maass, and L. de Mayer, *Progress in Reaction Kinetics*, Vol. 2, G. Porter, Ed., Pergamon Press, New York, 1964, p. 287.
3. L. M. Dorfman and M. S. Matheson, *Progress in Reaction Kinetics*, Vol. 3, G. Porter, Ed., Pergamon Press, New York, 1965, p. 237.
4. M. S. Matheson and J. Rabani, *J. Phys. Chem.*, **69**, 1324 (1965).
5. F. S. Dainton, *The Chemistry of the Electron*, Nobel Symposium 5, S. Claesson, Ed., Interscience Publishers, New York, 1967, p. 185.
6. C. Capellos and A. O. Allen, *J. Phys. Chem.*, **72**, 4265 (1968).
7. J. K. Thomas, *Rad. Res. Rev.*, **1**, 183 (1968).
8. C. Capellos, Technical Report 4054, Picatinny Arsenal, Dover, N.J., November 1970.

SYSTEM 37. COMPUTATION OF THE EXTENT OF DECAY OF TRANSIENTS DURING A PULSE

In fast reactions the rate of formation of a transient species has to be greater than the rate of its decay, otherwise it cannot be observed. If the rates of formation and decay are of comparable magnitude, steady-state conditions are approached rapidly and the concentration of the species might be so low that it is difficult if not impossible to measure.

As a result, techniques have been developed that permit the formation of relatively large concentrations of transient species in an extremely short time period. The formation of a transient under pulse radiolytic conditions can serve as an example. Modern pulsing machines can deliver a high intensity pulse of radiation (electrons, x-rays) of nanosecond duration.

Cases in which the length of the pulse is significantly long in comparison to the total decay time of the species formed, it becomes necessary to carry out corrections for decay during

System 37. Computation of the Extent of Decay of Transients

the pulse. The computation of this correction is given for species that disappear by a first- or by a second-order process.

Correction for a Fast Decay in a First-Order Process

The rate of formation of a transient species A during the pulse is given by the following differential equation:

$$\frac{d[A]}{dt} = GI - k[A] \tag{37-1}$$

$$= k\left[\frac{GI}{k} - [A]\right] \tag{37-2}$$

The $[A]$ is the concentration of the species at any time t during the pulse, G is the number of molecules of A formed per 100 eV energy absorbed by the system, I is the average intensity of radiation during the pulse expressed in terms of electron volts per liter; k is the first-order rate constant for the decay of A.

Equation 37-2 is integrated between the respective limits: $[A] = 0$ at the beginning of the pulse and $[A]_p$ its concentration observed at the end of the pulse; $t = 0$ at the beginning of the pulse and t_p denoting the pulse length. Hence

$$\frac{d[A]}{\frac{GI}{k} - [A]} = k\,dt \tag{37-3}$$

$$-\int_0^{[A]_p} \frac{d(GI/k - [A])}{(GI/k - [A])} = k\int_0^{t_p} dt \tag{37-4}$$

$$\ln\left[\frac{GI}{k} - [A]\right]\bigg|_0^{[A]_p} = -kt_p \tag{37-5}$$

$$\ln\left[\frac{GI/k}{(GI/k) - [A]_p}\right] = kt_p \tag{37-6}$$

$$\frac{GI/k}{(GI/k) - [A]_p} = e^{kt_p} \tag{37-7}$$

Thus $[A]_{\text{total}} = GIt_p$ is the concentration of the transient which would be present at the end of the pulse if no decay had occurred during the pulse. Combining this information with equation 37-7 yields

$$[A]_{\text{total}} = \left[\frac{[A]_p kt_p}{1 - e^{-kt_p}}\right] \tag{37-8}$$

Correction for a Fast Decay in a Second-Order Process

The rate of formation of a transient species A which is formed during a radiation pulse and which decays by second-order kinetics is given by the following differential equation:

$$\frac{d[A]}{dt} = GI - k[A]^2 \tag{37-9}$$

Upon rearrangement equation 37-9 can be integrated between limits: at the beginning of the pulse when $t = 0$, $[A] = 0$ and at the end of the pulse at t_p when $[A] = [A]_p$. Hence

$$\frac{d[A]}{[\sqrt{(GI/k)}]^2 - [A]^2} = k\, dt \tag{37-10}$$

$$\int_0^{[A]_p} \frac{d[A]}{[\sqrt{(GI/k)}]^2 - [A]^2} = k\int_0^{t_p} dt \tag{37-11}$$

$$\frac{\tan h^{-1}}{\sqrt{(GI/k)}} \frac{[A]}{\sqrt{(GI/k)}} \bigg|_0^{[A]_p} = kt \bigg|_0^{t_p} \tag{37-12}$$

$$\tan h^{-1}\left[\frac{[A]_p}{([A]_0/kt_p)^{1/2}}\right] = \tan h^{-1}\left([A]_p\left[\frac{t_p k}{[A]_0}\right]^{1/2}\right) \tag{37-13}$$

Hence

$$\tan h(kt_p[A]_0)^{1/2} = [A]_p(kt_p/[A]_0)^{1/2} \tag{37-14}$$

where $[A]_0$ is the concentration of A, if there has been no decay during its formation.

An approximation to equation 37-14 is

$$\frac{[A]_p}{[A]_0} = 1 - 0.313 kt_p[A]_p \qquad (37\text{-}15)$$

REFERENCES

1. C. Capellos and A. O. Allen, *J. Phys. Chem.*, **72**, 4265 (1968).
2. C. Capellos, Technical Report 4054, Picatinny Arsenal, Dover N.J., November 1970.

SYSTEM 38. EFFECT OF TEMPERATURE ON REACTION VELOCITY

From experimental observations it is known that chemical reaction velocities vary with temperature. The reason for this is that before a molecule A can react it has to become activated to A^*, that is, it has to acquire a certain minimum energy E_a called the activation energy.

$$A \xrightarrow{\text{slow}} A^* \xrightarrow{\text{fast}} \text{product}$$

This energy can be obtained by absorption of a certain quantity of heat.

If all molecules in a given system were to acquire this activation energy at the same rate, all reactions would be instantaneous. In fact only a small fraction of molecules manages to obtain the required energy. This fraction is given by the Boltzman factor $e^{-E_a/RT}$. The greater the energy required for activation, the fewer are the molecules possessing this energy, hence the slower is the reaction at a given temperature.

Reactions with Temperature Independent Activation Energies

If for a given reaction the activation energy is not itself temperature dependent, the variation of the rate constant with temperature is described by the Arrhenius equation

$$\frac{d \ln k}{dT} = \frac{E_a}{RT^2} \tag{38-1}$$

which yields on integration

$$\ln k = -\frac{E_a}{RT} + \ln A \tag{38-2}$$

or

$$k = Ae^{-E_a/RT} \tag{38-3}$$

A, the integration constant, is the frequency factor, for the present considered virtually temperature independent. A graphical presentation of the change in k with temperature (Fig. 1) illustrates the asymptotic approach of k to A, as the

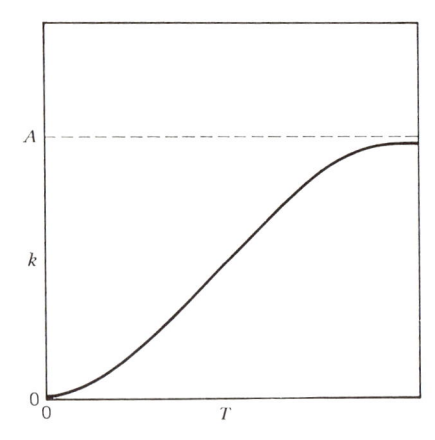

Figure 1

activation energy decreases toward zero. Very small or zero activation energies are encountered in reactions involving free radicals and atoms.

Integration of equation 38-1 between the appropriate limits gives

$$\ln \frac{k_2}{k_1} = -\frac{E_a}{R}\left[\frac{1}{T_2} - \frac{1}{T_1}\right] \tag{38-4}$$

System 38. Effect of Temperature on Reaction Velocity

If the activation energy and the value of the rate constant at one temperature are known, the rate constant for any other temperature can be calculated.

Since it is difficult to predict the activation energies of chemical reactions, they must be measured experimentally. This is done by measuring k at several temperatures and plotting $\ln k$ against $1/T$. If E_a and A are indeed temperature independent, the plot should give a straight line with a slope equal to $-E_a/R$.

Temperature Dependence of the Activation Energy

Reactions with small or zero activation energies show a nonlinear temperature dependence since in such reactions the A factor in the Arrhenius equation becomes a function of temperature itself. For a detailed discussion of a more rigorous treatment the reader should check references (1–8).

A relatively accurate equation for the treatment of experimental data is

$$k = AT^n e^{-E/RT} \tag{38-5}$$

or

$$\ln k = -\frac{E}{RT} + n \ln T + \ln A \tag{38-6}$$

where A is a constant, and n is an exponent whose value depends on the choice of theoretical treatment as well as on the type and nature of the reaction under consideration.

Differentiation of (38-6) with respect to time gives an equation equivalent to the Arrhenius relation if it is assumed that E is constant:

$$\frac{d \ln k}{dt} = \frac{n}{T} + \frac{E}{RT^2} = \frac{E_a}{RT^2} \tag{38-7}$$

hence

$$E = E_a - nRT \tag{38-8}$$

The term nRT has such a small value that in most experimental work E will appear constant as the temperature varies.

REFERENCES

1. S. Arrhenius, *Z. Phys. Chem.*, **4**, 226 (1889).
2. H. Goldschmidt, *Phys. Z.*, **10**, 206 (1909).
3. Tolman, *Statistical Mechanics*, Chemical Catalog Co., New York, 1927.
4. V. K. LaMer, *J. Chem. Phys.*, **1**, 289 (1933).
5. V. K. LaMer, *J. Am. Chem. Soc.*, **55**, 1739 (1933).
6. E. A. Moelwyn-Hughes, *The Kinetics of Reaction in Solution*, 2nd ed., Oxford University Press, London, 1942, Chap. 2.
7. W. S. Horton, *J. Phys. Colloid Chem.*, **52**, 1129 (1948).
8. S. Benson, *Thermochemical Kinetics*, John Wiley & Sons, Inc., New York, 1968.

SYSTEM 39. EFFECT OF PRESSURE ON REACTION VELOCITY

A general theory of the effect of pressure on the velocity of chemical reactions was first proposed by Van't Hoff in 1901. According to this theory, the equilibrium constant $K^{\ddagger}$ for the reaction

$$A + B \underset{k_2}{\overset{k_1}{\rightleftharpoons}} AB^{\ddagger}$$

(where $AB^{\ddagger}$ is an activated complex) is related to the free energy change $\Delta G^{\ddagger}$ by the thermodynamic relationship

$$\ln K^{\ddagger} = -\frac{\Delta G^{\ddagger}}{RT} \qquad (39\text{-}1)$$

where R is the gas constant and T the temperature in °K. Combining this relation with the equation relating internal energy (E), pressure (P), and entropy (S) with change in free energy one obtains an expression which can be differentiated

System 39. Effect of Pressure on Reaction Velocity

with respect to P at constant temperature:

$$\Delta G^{\ddagger} = \Delta E^{\ddagger} + P\,\Delta V^{\ddagger} - T\,\Delta S^{\ddagger} \tag{39-2}$$

$$\ln K^{\ddagger} = -\frac{\Delta E^{\ddagger}}{RT} + \frac{\Delta S^{\ddagger}}{R} - P\,\Delta V^{\ddagger} \tag{39-3}$$

$$\left(\frac{\partial \ln K^{\ddagger}}{\partial P}\right)_T = -\frac{\Delta V^{\ddagger}}{RT} \tag{39-4}$$

The Δ operator indicates the difference in the particular molar property, each in a defined reference state, between complex $AB^{\ddagger}$ and reactants.

To find the relationship between k_1 and $\Delta V^{\ddagger}$ use is made of the rate theory, which postulates that k_1 is related to the equilibrium constant $K^{\ddagger}$ by the general frequency factor kT/h

$$k_1 = \frac{kT}{h} K^{\ddagger} \tag{39-5}$$

The term k (Boltzman constant) allows for the probability that an activated molecule will give the products of the reaction; h is Planck's constant and T temperature. The $K^{\ddagger}$ in equation 39-5 is not an equilibrium constant in the true sense, but only similar to one. By taking the logarithm of (39-5) and differentiating it with respect to pressure at constant temperature, one obtains an expression which gives $\Delta V^{\ddagger}$ in terms of k_1 and P when combined with equation 39-4:

$$\ln k_1 = \ln\frac{kT}{h} + \ln K^{\ddagger} \tag{39-6}$$

$$\left(\frac{\partial \ln k_1}{\partial P}\right)_T = \left(\frac{\partial \ln K^{\ddagger}}{\partial P}\right)_T \tag{39-7}$$

Hence

$$\left(\frac{\partial \ln k_1}{\partial P}\right)_T = -\left(\frac{\Delta V^{\ddagger}}{RT}\right) \tag{39-8}$$

Equation 39-8 implies that the rate constant of a chemical reaction increases with increasing pressure, when the volume of the activated complex $AB^{\ddagger}$ is less than the volume of the reactants A and B.

If the pressure effect on k_1 is independent of temperature, equation 39-8 can be integrated between the corresponding limits to yield

$$\int_{(k_1)_{p_1}}^{(k_1)_{p_2}} d\ln k_1 = -\left(\frac{\Delta V^{\ddagger}}{RT}\right) \int_{p_1}^{p_2} dP \qquad (39\text{-}9)$$

$$\ln \left[\frac{(k_1)_{p_2}}{(k_1)_{p_1}}\right] = -\left(\frac{\Delta V^{\ddagger}}{RT}\right)(P_2 - P_1) \qquad (39\text{-}10)$$

where $(k_1)_{p_1}$ and $(k_1)_{p_2}$ are the rate constants at pressure P_1 and P_2. Substitution of typical experimental values into this equation shows that thousands of atmospheres of pressure are required to cause a change in $\Delta V^{\ddagger}$.

In general $\Delta V^{\ddagger}$ is determined experimentally by measuring k_1 at various pressures and ploting $\log k_1$ against pressure. A linear plot indicates that volume change due to formation of an active complex remains constant with change in pressure. From the slope of such a linear plot one can compute $\Delta V^{\ddagger}$:

$$\Delta V^{\ddagger} = \text{slope} \times 2.303 \times RT \qquad (39\text{-}11)$$

REFERENCES

1. V. Rothmund, *Z. Phys. Chem.*, **20**, 168 (1896).
2. E. W. Fawcett and C. H. Gibson, *J. Chem. Soc.*, 386 (1934).
3. C. H. Gibson, E. W. Fawcett, and M. W. Perrin, *Proc. Roy. Soc.*, **A150**, 223 (1935).
4. M. G. Evans and M. Polanyi, *Trans. Faraday Soc.*, **31**, 875 (1935).
5. G. M. Evans and M. Polanyi, *Trans. Faraday Soc.*, **32**, 1333 (1936).
6. R. R. Williams, M. W. Perrin, and C. H. Gibson, *Proc. Roy. Soc.*, **A154**, 684 (1936).

7. M. W. Perrin, *Trans. Faraday Soc.*, **34**, 144 (1938).
8. D. M. Newitt and A. Wassermann, *Trans. Chem. Soc.*, 735 (1940).
9. S. D. Hamann and D. R. Teplitzky, *Disc. Faraday Soc.*, **22**, 114 (1956).
10. S. D. Hamann, *Physico-Chemical Effects of Pressure*, Butterworth and Co. Ltd., London, 1957.

SYSTEM 40. EFFECT OF IONIC STRENGTH ON VELOCITY OF IONIC REACTIONS

$$A^{Z_a} + B^{Z_b} \underset{k_2}{\overset{k_1}{\rightleftharpoons}} AB^{(Z_a+Z_b)} \overset{k_3}{\longrightarrow} C \tag{I}$$

This reaction between two ionic species A^{Z_a} and B^{Z_b} proceeds through a transition-state complex $AB^{(Z_a+Z_b)}$ which is in equilibrium with the reactants. The equilibrium properties of such reactions can be greatly effected by other ionic species which may be present in addition to the reactants.

The variable that determines the effect of ions on the equilibrium is the ionic strength, defined by

$$\mu = \tfrac{1}{2} \Sigma\, m_i z_i^2 \tag{40-1}$$

where m_i is the molality (in moles of solute per 1000 gram of solvent) and z_i the charge of the *i*th ionic species present in the system.

The effect of electrostatic interactions of ionic species can be successfully treated by the activity rate theory, which was developed by Bronsted, Bjerrum, Debye, and Huckel.

The equilibrium of the transition-state complex is not affected by the presence of charged species if the rate of chemical transformation is smaller than the rate of dissociation of the complex ($k_3 \ll k_1$).

$$K^{\ddagger} = \frac{\mathbf{a}_{AB^{\ddagger}}}{\mathbf{a}_A \mathbf{b}_B} = \frac{[AB^{\ddagger}]}{[A][B]} \frac{\gamma_{AB^{\ddagger}}}{\gamma_A \gamma_B} \tag{40-2}$$

In dealing with ions, the equilibrium must be expressed in terms of activities **a** and the corresponding activity coefficients γ.

120   Miscellaneous

The concentration of the activated complex is

$$[AB^{\ddagger}] = [A][B]K^{\ddagger}\frac{\gamma_A\gamma_B}{\gamma_{AB^{\ddagger}}} \quad (40\text{-}3)$$

The rate of reaction, in terms of the absolute rate theory can be written as

$$-\frac{d[A]}{dt} = k_3[A][B] = \left[\frac{kT}{h}\right][AB^{\ddagger}] \quad (40\text{-}4)$$

$$k_3 = \left[\frac{kT}{h}\right]\frac{[AB^{\ddagger}]}{[A][B]} \quad (40\text{-}5)$$

where k is the Boltzman constant, T the absolute temperature, and h Planck's constant.

Combining equations 40-3 and 40-5 yields

$$k_3 = \left[\frac{kT}{h}\right]\left[\frac{\gamma_A\gamma_B}{\gamma_{AB^{\ddagger}}}\right]K^{\ddagger} \quad (40\text{-}6)$$

or

$$\log k_3 = \log\left[\frac{kTK^{\ddagger}}{h}\right] + \log\gamma_A + \log\gamma_B - \log\gamma_{AB^{\ddagger}} \quad (40\text{-}7)$$

According to the expanded Debye-Huckel theory, the mean activity coefficient γ_i is related to the ionic strength μ by the following equation:

$$\log\gamma_{i(\pm)} = \frac{-A|Z_+Z_-|\sqrt{\mu}}{1 + Bd\sqrt{\mu}} \quad (40\text{-}8)$$

where A and B are constants and d is the average effective diameter of the ions. Combining (40-7) and (40-8) yields

$$\log k_3 = \log\left[\frac{kTK^{\ddagger}}{h}\right] + [-AZ_A^2 - AZ_B^2 + A(Z_A + Z_B)^2]$$

$$\times \frac{\sqrt{\mu}}{1 + Bd\sqrt{\mu}} \quad (40\text{-}9)$$

hence

$$\log k_3 = \log k_0 + \frac{A2Z_AZ_B\sqrt{\mu}}{1 + Bd\sqrt{\mu}} \quad (40\text{-}10)$$

System 40. Effect of Ionic Strength on Velocity

where

$$k_0 = \left[\frac{kT}{h}\right] K^{\ddagger}$$

represents the rate constant at infinite dilution.

For aqueous solutions at 25°C, $A = 0.509$, $B = 0.329 \times 10^8$ and d is usually of the order of a few Angstroms. Hence for aqueous solutions the change in rate constants with ionic strength is given by (40-11), known as the Brønsted equation:

$$\log\left(\frac{k_3}{k_0}\right) = \left[\frac{1.018 Z_A Z_B \sqrt{\mu}}{1 + 0.329 d \sqrt{\mu}}\right] \tag{40-11}$$

A plot of $\log k_3$ versus $\sqrt{\mu}$ should give a straight line with a slope nearly equal to the product of the ionic charges. The change in k_3 with ionic strength is known as the primary salt effect. The following three cases can occur:

Product of $Z_A Z_B$	k_3
+	Increases with ionic strength
−	Decreases with ionic strength
0	Independent of ionic strength

The change of the reaction velocity with ionic strength is illustrated on several classical reactions, (I), (II), (III), (IV), and (V) in Fig. 2. It should be remembered that it is the magnitude and sign of $Z_A Z_B$ for the rate determining step, which controls the rate of the reaction, for example:

$$2[Co(NH_3)_5 Br]^{++} + Hg^{++}$$
$$+ 2H_2O \rightarrow 2[Co(NH_3)_5 H_2O]^{+++} + HgBr_2 \quad \text{(I)}$$
$$Z_A \times Z_B = +2 \times +2 = +4$$
$$S_2O_8^{=} + 2I^- \rightarrow I_2 + 2SO_4^- \quad \text{(II)}$$
$$Z_A \times Z_B = -2 \times -1 = +2$$
$$CH_3COOC_2H_5 + OH^- \rightarrow CH_3COO^- + C_2H_5OH$$
$$Z_A \times Z_B = 0 \times -1 = 0 \quad \text{(III)}$$

Figure 2

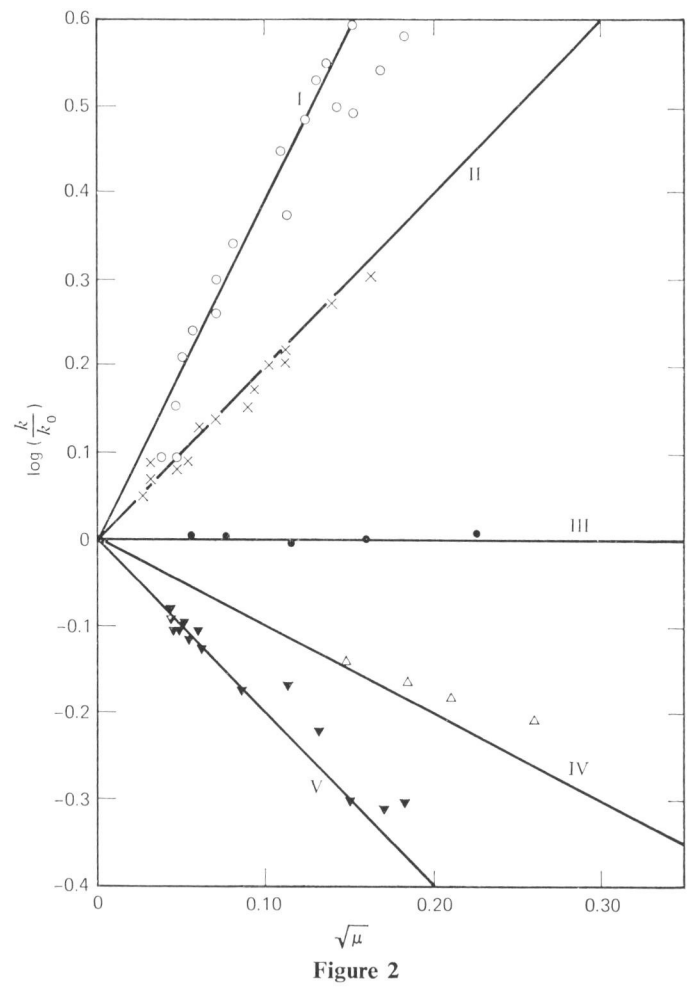

$$H_2O_2 + 2H^+ + 2Br^- \rightarrow 2H_2O + Br_2$$
$$Z_A \times Z_B = +1 \times -1 = -1 \qquad (IV)$$
$$[Co(NH_3)_5Br]^{++} + OH^- \rightarrow [Co(NH_3)_5OH]^{++} + Br^-$$
$$Z_A \times Z_B = +2 \times -1 = -2 \qquad (V)$$

REFERENCES

1. J. N. Brønsted, *Z. Phys. Chem.*, **102**, 169 (1922).
2. N. Bjerrum, *Z. Phys. Chem.*, **108**, 82 (1924).
3. J. A. Christiansen, *Z. Phys. Chem.*, **113**, 35 (1924).
4. J. N. Brønsted, *Z. Phys. Chem.*, **115**, 337 (1925).
5. N. Bjerrum, *Z. Phys. Chem.*, **118**, 251 (1925).
6. P. Debye and J. McAulay, *Phys. Z.*, **26**, 22 (1925).
7. E. Huckel, *Phys. Z.*, **26**, 93 (1925).
8. V. R. Livingston, *J. Am. Chem. Soc.*, **48**, 53 (1926).
9. J. N. Brønsted and V. R. Livingston, *J. Am. Chem. Soc.*, **49**, 435 (1927).
10. V. K. LaMer, *J. Am. Chem. Soc.*, **51**, 334 (1929).
11. C. V. King and M. B. Jacobs, *J. Am. Chem. Soc.*, **53**, 1704 (1931).
12. G. Scatchard, *Chem. Rev.*, **10**, 229 (1932).
13. V. K. LaMer, *Chem. Rev.*, **10**, 179 (1932).
14. K. J. Laidler and H. Eyring, *Ann. N.Y. Acad. Sci.*, **39**, 303 (1940).
15. R. P. Bell and J. E. Prue, *J. Chem. Soc.*, 362 (1949).
16. C. W. Davies, *Progress in Reaction Kinetics*, Vol. I., G. Porter, Ed., Pergamon Press, New York, 1961, p. 161.
17. H. A. Schwarz and G. Czapski, *J. Phys. Chem.*, **66**, 471 (1962).
18. E. Collinson, F. S. Dainton, D. R. Smith, and S. Tazuke, *Proc. Chem. Soc.*, 140, 1962.

Problems

Problem 1

Using the method of isolation, show how you would determine the order of reaction for the individual species taking part in a reaction described by the following differential equation:

$$-\frac{dA}{dt} = k[A][B][C]^2$$

Problem 2

Consider the following mechanism

$$A \xrightarrow{k} B$$

At $t = 0$ $[A]_0$ $[B]_0$
At $t = t$ $[A]$ $[B]$

Derive the expression $[B] = [A]_0(2 - e^{-k_1 t})$, and discuss the conditions under which this relationship holds true.

Problem 3

$$2A \xrightarrow{k} B$$

At $t = 0$ $[A]_0$ $[B]_0$
At $t = t$ $[A]$ $[B]$

(a) Assuming the mechanism and experimental conditions above derive the following expression for [B]

$$[B] = [B]_0 + [A]_0\left[1 - \frac{1}{1 + 2k[A]_0 t}\right]$$

(b) Discuss assumptions and experimental conditions under which the following expression holds true

$$[B] = 2[A]_0\left[1 + \frac{k_1[A]_0 t}{1 + 2k[A]_0 t}\right]$$

Problem 4

Consider the mechanism

$$A + B \xrightarrow{k} C$$

At $t = 0$ $[A]_0$ $[B]_0$ $[C]_0$
At $t = t$ $[A]$ $[B]$ $[C]$

Prove that for this reaction

$$[C] = [C]_0 + \left(\frac{[A]_0[B]_0[1 - e^{-k([B]_0 - [A]_0)t}]}{[B]_0 - [A]_0[e^{-k'([B]_0 - [A]_0)t}]}\right)$$

Problem 5

Consider the system:

$$A + A + A \xrightarrow{k} \text{products}$$

and prove that

$$[A] = \left[\frac{[A]_0}{(1 + 2k[A]_0^2 t)^{1/2}}\right]$$

The $[A]_0$ is the initial concentration of reactant A.

Problem 6

	$2A + B \xrightarrow{k} C$		
At $t = 0$	$[A]_0$	$[B]_0$	0
At $t = t$	$[A]$	$[B]$	$[C]$

Show that (1)

$$\ln\left[\frac{([B]_0/[B])}{([A]_0/[A])}\right] = \left[\frac{[A]_0}{[A]} - 1\right]\left[2\frac{[B]_0}{[A]_0} - 1\right] - k(2[B]_0 - [A]_0)^2 t$$

and (2)

$$[C] = \frac{[A]_0[A]}{2(2[B]_0 - [A]_0)}\left[(2[B]_0 - [A]_0)^2 kt + \ln\left(\frac{[A]_0[B]}{[B]_0[A]}\right)\right]$$

Assuming that $[A]_0$ and $[B]_0$ are known and that $[A]$ and $[B]$ can be determined as a function of time, show that k can be determined graphically.

Problem 7

Consider the mechanism

	$A \xrightarrow{k_1} B \xrightarrow{k_2} C$		
At $t = 0$	$[A]_0$	0	0
At $t = t$	$[A]$	$[B]$	$[C]$

Prove that

$$k_1 = \frac{1}{t}\ln\left[1 + \frac{[B]}{[A]} + \frac{[C]}{[A]}\right]$$

Problem 8

Consider the case of first-order opposing reactions:

$$A \underset{k_2}{\overset{k_1}{\rightleftharpoons}} B$$

Problems

The initial concentrations ($t = 0$) are $[A]_0 = a$ and $[B]_0 = 0$. At time t, $[A] = (a - x)$ and $[B] = x$.

1. Set up the differential rate equation.
2. Prove that in general

$$(k_1 + k_2)t = \ln\left[\frac{k_1 a}{(k_1 + k_2(a - x) - k_2 a}\right]$$

3. Under what conditions is

$$\left[\frac{(a - x)}{a}\right] = \left[\frac{k_2}{(k_1 + k_2)}\right]$$

4. Prove that at equilibrium when $x = x_e$

$$(k_1 + k_2)t = \ln\left[\frac{x_e}{x_e - x}\right]$$

Problem 9

Consider the system

$$A \underset{k_2}{\overset{k_1}{\rightleftarrows}} B$$

At $t = 0$ $[A]_0$ $[B]_0$
At $t = t$ $[A]$ $[A]$

If the relationship $[B] - [B]_0 = [A]_0 - [A]$ holds between the concentrations at time 0 and t, derive the following expressions:

1. $A = \left[\dfrac{([B]_0 + [A]_0)}{(1 + k_1/k_2)}\right] - \left[\dfrac{[B]_0 - (k_1/k_2)[A]_0}{(1 + k_1/k_2)}\right]e^{-(k_1+k_2)t}$

2. $B = \left[\dfrac{([B]_0 + [A]_0)}{(1 + k_2/k_1)}\right] - \left[\dfrac{[A]_0 - (k_2/k_1)[B]_0}{(1 + k_2/k_1)}\right]e^{-(k_1+k_2)t}$

Problem 10

$$A + B \underset{k_2}{\overset{k_1}{\rightleftharpoons}} C + D$$

At $t = 0$ $[A]_0$ $[B]_0$ 0   0

At $t = t$ $([A]_0 - x)$ $([B]_0 - x)$ x x

Show that the rate constant k_1 can be expressed by the following equation:

$$k_1 = \frac{x_e}{t[2[A]_0[B]_0 - x_e([A]_0 + [B]_0)]} \ln\left[\frac{[A]_0[B]_0(x_e + x) - x_e([A]_0 + [B]_0)x}{[A]_0[B]_0(x_e - x)}\right]$$

where x_e is the value for x at equilibrium.

Problem 11

Consider the mechanism

$$A + C \xrightarrow{k_1} \text{product}_1$$
$$B + C \xrightarrow{k_2} \text{product}_2$$

Assuming that $k_1 = k_2 = k$ and that the initial concentrations of the reactants are $[C]_0 = [A]_0 + [B]_0$, show that

$$[A] + [B] = \left[\frac{[A]_0 + [B]_0}{1 + k([A]_0 + [B]_0)t}\right]$$

Problem 12

$$2A \xrightarrow{k} \text{products}$$

Assuming that the reaction above takes place under conditions of constant temperature and volume and that the concentration $[A]_t$ of reactant A is given in terms of a measurable property

$$w_t = Me^{q[A]_t}$$

where M and q are constants that depend on temperature only, and t is time. Prove that the following relationship is true:

$$k = \frac{q}{2t}\left[\frac{\ln(w_0/w_t)}{(\ln(w_t/M))(\ln(w_0/M))}\right]$$

Appendix I

DIFFERENTIALS

1. $\dfrac{d(a)}{dx} = 0$, where a is a constant

2. $\dfrac{d(x)}{dx} = 1$

3. $\dfrac{d(ax)}{dx} = a$

4. $\dfrac{d(au)}{dx} = a\dfrac{du}{dx}$

5. $\dfrac{d}{dx}(u + v - w) = \dfrac{du}{dx} + \dfrac{dv}{dx} - \dfrac{dw}{dx}$

6. $\dfrac{d}{dx}(uv) = u\dfrac{dv}{dx} + v\dfrac{du}{dx}$

7. $\dfrac{d}{dx}\left(\dfrac{u}{v}\right) = \left[\dfrac{v\dfrac{du}{dx} - u\dfrac{dv}{dx}}{v^2}\right]$

8. $\dfrac{d}{dx}(u^n) = nu^{n-1}\dfrac{du}{dx}$

9. $\dfrac{d}{dx}(\ln u) = \dfrac{1}{u}\dfrac{du}{dx}$

10. $\dfrac{d}{dx}(e^u) = e^u\dfrac{du}{dx}$

INDEFINITE INTEGRALS

1. $\displaystyle\int a\,dx = ax$

2. $\displaystyle\int a f(x)\,dx = a \int f(x)\,dx$

3. $\displaystyle\int f(y)\,dx = \int \frac{f(y)}{y'}\,dy$ where $y' = dy/dx$

4. $\displaystyle\int (u + v)\,dx = \int u\,dx + \int u\,dx$ where u and v are functions of x

5. $\displaystyle\int u\,dv = u\int dv - \int v\,du = uv - \int v\,du$

6. $\displaystyle\int x^n\,dx = \left(\frac{x^{n+1}}{n+1}\right)$ for $n \neq -1$

7. $\displaystyle\int \frac{f'(x)\,dx}{f(x)} = \log f(x)$

8. $\displaystyle\int \frac{dx}{x} = \log x$

9. $\displaystyle\int e^x\,dx = e^x$

10. $\displaystyle\int e^{ax}\,dx = \frac{e^{ax}}{a}$

11. $\displaystyle\int \frac{dx}{a^2 + x^2} = \frac{1}{a}\tan^{-1}\left[\frac{x}{a}\right]$

12. $\displaystyle\int \frac{dx}{a^2 - x^2} = \frac{1}{a}\tan h^{-1}\left[\frac{x}{a}\right]$ for $a^2 > x^2$

13. $\int (a + bx)^n \, dx = \dfrac{(a + bx)^{n+1}}{(n + 1)b}$ for $n \neq -1$

14. $\int \dfrac{e^{ax} \, dx}{x} = \log x + \dfrac{ax}{1!} + \dfrac{a^2 x^2}{2 \cdot 2!} + \dfrac{a^3 x^3}{3 \cdot 3!} + \cdots$

DEFINITE INTEGRALS

1. $\int_a^b f(x) \, dx = F(b) - F(a)$. If $f(x)$ is continuous in (a, b) and $F(x)$ is a function such that: $F'(x) = f(x)$

2. $\int_a^b A f(x) \, dx = A \int_a^b f(x) \, dx$, where A represents a constant

3. $\int_a^b [f(x) \pm g(x)] \, dx = \int_a^b f(x) \, dx \pm \int_a^b g(x) \, dx$

DEFINITE INTEGRALS OF SOME SPECIAL FUNCTIONS

1. $\int_a^b du = u \Big|_a^b = b - a$

2. $\int_a^b \dfrac{du}{u} = \ln |u| \Big|_a^b = \ln (b) - \ln (a)$

3. $\int_a^b u^n \, du = \dfrac{u^{n+1}}{n + 1} \Big|_a^b = \dfrac{b^{n+1}}{n + 1} - \dfrac{a^{n+1}}{n + 1}$ for $n \neq -1$

4. $\int_a^b e^u \, du = e^u \Big|_a^b = e^b - e^a$

Appendix II

GUIDE LINE TO THE OPERATOR METHOD

The operator method can be used for solving linear differential equations that describe kinetic processes in certain chemical systems, provided the prerequisite rules are observed. Here, only a few simple examples are given to illustrate the method. For a more rigorous treatment the reader should consult standard mathematical texts.

Case A requires the specific initial conditions, where the function $f(t)$ and all its derivatives with respect to time are equal to zero when $t = 0$.

Example I

$$\frac{dx}{dt} + ax = 1 \qquad (1)$$

The a = constant and x, t = variables. Set $d/dt = S$, and treat S the operator as a constant. Substitution of S into (1) gives

$$Sx + ax = 1 \qquad (2)$$

or

$$x = \frac{1}{S + a} \qquad (3)$$

The right-hand side of (3) is called a transform. Transforms and their corresponding functions of time called originals are listed in mathematical tables. In this case the original, which describes x as a function of time, is

$$x = \frac{1}{a}\left[1 - e^{-at}\right] \qquad (4)$$

Example II

$$\frac{d^2x}{dt^2} + 2a\frac{dx}{dt} + a^2x = 1 \qquad (1)$$

Since the operator S is independent of x,

$$\frac{d^2x}{dt^2} = \frac{d}{dt}\left[\frac{dx}{dt}\right] = \frac{d}{dt}Sx = S^2x \qquad (2)$$

and

$$S^2x + 2aSx + a^2x = 1 \qquad (3)$$

$$x(S^2 + 2aS + a^2) = 1 \qquad (4)$$

$$x = \frac{1}{(S^2 + 2aS + a^2)} = \frac{1}{(S + a)^2} \qquad (5)$$

Substituting the original for the transform gives the final solution

$$x = \frac{1}{a^2} - \left[\frac{1 + at}{a^2}\right]e^{-at} \qquad (6)$$

Case B deals with the situation where the initial conditions for x are not zero, when $t = 0$. In this case substitution of S for derivatives is carried out as follows:

$$\frac{dx}{dt} = Sx - Sx_0 \qquad (1)$$

$$\frac{d^2x}{dt^2} = S^2x - (S^2x_0 - Sx_1) \qquad (2)$$

Appendix II

and so on where

$$x_0 = f(t) \quad \text{at} \quad t = 0 \tag{3}$$

$$x_1 = \frac{df(t)}{dt} \quad \text{at} \quad t = 0 \tag{4}$$

$$x_2 = \frac{d^2f(t)}{dt^2} \quad \text{at} \quad t = 0 \tag{5}$$

and so on.

Example

$$\frac{dx}{dt} + ax = 1 \tag{6}$$

Substitution for the derivative under the conditions above gives

$$Sx - Sx_0 + ax = 1 \tag{7}$$

Assuming that $x_0 = 1$, equation 7 reduces to

$$x = \frac{1}{S + a} + \frac{S}{S + a} \tag{8}$$

Replacing the transforms by the originals yields the answer to (6)

$$x = \left[\frac{1}{a}\right](1 - e^{-at}) + e^{-at} \tag{9}$$

TABLE OF TRANSFORMS AND ORIGINALS

Transforms	Originals
$\dfrac{1}{S}$	t
$\dfrac{1}{S^n}$	$\dfrac{t^n}{n!}$
$\dfrac{1}{(S \pm a)}$	$\left[\dfrac{1}{\pm a}\right] - \left[\dfrac{1}{\pm a}\right] e^{\mp at}$
$\dfrac{S}{(S \pm a)}$	$e^{\mp at}$
$\dfrac{(S+b)}{(S+a)}$	$\left[\dfrac{b}{a}\right] - \left[\dfrac{(b-a)}{a}\right] e^{-at}$
$\dfrac{1}{(S+a_1)(S+a_2)}$	$\left[\dfrac{1}{a_1 a_2}\right] - \left[\dfrac{1}{a_1(a_2 - a_1)}\right] e^{-a_1 t}$
	$\quad - \left[\dfrac{1}{a_2(a_1 - a_2)}\right] e^{-a_2 t}$
$\dfrac{S}{(S+a_1)(S+a_2)}$	$\left[\dfrac{1}{(a_2 - a_1)}\right] e^{-a_1 t} + \left[\dfrac{1}{(a_1 - a_2)}\right] e^{-a_2 t}$
$\dfrac{(S+b)}{(S+a_1)(S+a_2)}$	$\left[\dfrac{b}{a_1 a_2}\right] - \left[\dfrac{(b-a_1)}{a_1(a_2 - a_1)}\right] e^{-a_1 t}$
	$\quad - \left[\dfrac{(b-a_2)}{a_2(a_1 - a_2)}\right] e^{-a_2 t}$
$\dfrac{1}{(S+a_1)(S+a_2)(S+a_3)\cdots(S+a_n)}$	$\dfrac{1}{a_1 a_2 a_3 \cdots a_n} +$
	$\quad - \dfrac{1}{a_1(a_2 - a_1)(a_3 - a_1)(a_4 - a_1)\cdots(a_n - a_1)} e^{-a_1 t}$
	$\quad - \dfrac{1}{a_2(a_1 - a_2)(a_3 - a_2)(a_4 - a_3)\cdots(a_n - a_2)} e^{-a_2 t} - \cdots$
	$\quad - \dfrac{1}{a_n(a_1 - a_n)(a_2 - a_n)(a_3 - a_n)\cdots(a_{n-1} - a_n)} e^{-a_n t}$

Appendix II

TABLE OF TRANSFORMS AND ORIGINALS (*Contd*).

Transforms	Originals

$$\frac{S}{(S + a_1)(S + a_2)(S + a_3) \cdots (S + a_n)}$$

$$\frac{1}{(a_2 - a_1)(a_3 - a_1) \cdots (a_n - a_1)} e^{-a_1 t}$$
$$+ \frac{1}{(a_1 - a_2)(a_3 - a_2) \cdots (a_n - a_2)} e^{-a_2 t}$$
$$+ \frac{1}{(a_1 - a_3)(a_2 - a_3) \cdots (a_n - a_3)} e^{-a_3 t} + \cdots$$
$$+ \frac{1}{(a_1 - a_n)(a_2 - a_n) \cdots (a_{n-1} - a_n)} e^{-a_n t}$$

$$\frac{S + b}{(S + a_1)(S + a_2)(S + a_3) \cdots (S + a_n)}$$

$$\frac{b}{a_1 a_2 a_3 \cdots a_n} - \frac{b - a_1}{(a_2 - a_1)(a_3 - a_1)(a_4 - a_1) \cdots (a_n - a_1)} e^{-a_1 t}$$
$$- \frac{b - a_2}{(a_1 - a_2)(a_3 - a_2)(a_4 - a_2) \cdots (a_n - a_2)} e^{-a_2 t} - \cdots$$
$$- \frac{b - a_n}{(a_1 - a_n)(a_2 - a_n)(a_3 - a_n) \cdots (a_{n-1} - a_n)} e^{-a_n t}$$

$$\frac{S(S + b)}{(S + a_1)(S + a_2)(S + a_3) \cdots (S + a_n)}$$

$$\frac{b - a_1}{(a_2 - a_1)(a_3 - a_1) \cdots (a_n - a_1)} e^{-a_1 t}$$
$$+ \frac{b - a_2}{(a_1 - a_2)(a_3 - a_2) \cdots (a_n - a_2)} e^{-a_2 t} +$$
$$\frac{b - a_3}{(a_1 - a_3)(a_2 - a_3) \cdots (a_n - a_3)} e^{-a_3 t} + \cdots$$
$$+ \frac{b - a_n}{(a_1 - a_n)(a_2 - a_n) \cdots (a_{n-1} - a_n)} e^{-a_n t}$$